HOUSE INSPECTOR

Duncan Marshall

BSc, MCIOB, MIMBM
Senior Lecturer

Nigel Dann

BSc,(Hons), MIOC
Senior Lecturer

Faculty of the Built Environment
UWE, Bristol

2005

Routledge
Taylor & Francis Group

LONDON AND NEW YORK

First published 2005 by Estates Gazette

Published 2014 by Routledge
2 Park Square, Milton Park, Abingdon, Oxon OX14 4RN
711 Third Avenue, New York, NY 10017, USA

*Routledge is an imprint of the Taylor & Francis Group,
an informa business*

ISBN 13: 978-0-7282-0489-8 (pbk)

Design and Typesetting by Ted Masters

INTRODUCTION

Our aim in writing this book was to produce a concise yet comprehensive guide to building defects and building inspection. There are many text books in this area but most of them are aimed at professional building surveyors carrying out detailed surveys on behalf of potential house buyers. We felt that there was a need for a more general and accessible book, aimed at the thousands of people who buy, sell or manage property, which briefly explains the construction of new and old houses and which also provides a series of elemental checklists to ensure that any significant defects do not go unnoticed.

If you are a trainee or general practice surveyor, a maintenance inspector, a Home Inspector, a housing manager, an estate agent, a planner, or even if you are a private purchaser or investor, this book should improve your knowledge and understanding of potential problems and provide a simple framework for a competent building inspection. Indeed, if you are selling a house and want to understand or even challenge the recently introduced Home Inspection reports you should find this book useful.

Most chapters include three main elements:
- A brief history of building over the last hundred years or so to help you identify and understand specific construction details.
- Illustrations and explanations of the most common building defects.
- A simple checklist to help you carry out a systematic and thorough inspection.

We have included chapters on all the major construction elements, (floors, walls, roofs, finishes, heating, drainage etc) and chapters on specific problems such as dampness, condensation and rot. The book is illustrated with hundreds of photos and line drawings, nearly all of them in colour. So, whether, as a practitioner, you are employed in buying, selling, managing or maintaining houses or whether, as a layperson, you are buying a property to develop or live in, this book should help you make sound decisions and, possibly, avoid costly mistakes. *Duncan Marshall and Nigel Dann*

INTRODUCTION

In the UK, there are three basic foundation types:

• Strip foundations • Piled foundations • Raft foundations

Strip foundations are by far the most common. The foundation is basically a strip, or ribbon, of insitu concrete running under all the loadbearing walls. These will normally include all the external walls and possibly some, or all, of the internal walls. The depth and width of the strip depends on the building load and the nature of the ground. In many cases, these foundations do not need specialist design; the foundation size can be determined by referring to tables in the Building Regulations.

Piled foundations can be of various types. They transmit the loads from the building through weak, compressible, or unstable strata to firmer ground beneath (end-bearing piles). In clay and other cohesive soils, piles can be used to distribute the loads into the ground through the friction forces along the length of the pile sides (friction piles). Piles are usually made from insitu or pre-cast concrete but also can be steel and timber. In housing built from loadbearing brickwork, a reinforced concrete beam bridges the piles and directly supports the walls.

Rafts are an expensive form of construction, probably the most costly of the three, and are used where only a very low load can be applied, for example, on soft or variable ground. They are also used where differential settlement is likely or where there is a risk of subsidence (they are common in mining areas). The raft is a rigid slab of concrete, reinforced with steel, which spreads the building load over the whole ground floor area.

Early foundations

Prior to the end of the 19th century it was not uncommon to build external walls on a rubble foundation or even on no foundation at all. Many of these houses still stand today.

At the end of the 19th century building control became increasingly onerous, especially in London. The 1894 London Building Act specified foundation and wall sizes for a variety of wall heights and lengths. In the provinces, bye-laws were often much less demanding.

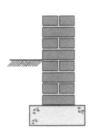

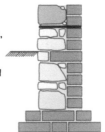

By the early 1900s most houses were built with concrete or brick footings. In some cases a brick footing might sit on a concrete base.

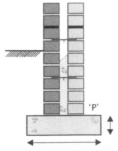

Model bye-laws in the late 1930s required a concrete foundation with a width equal to or greater than twice the wall and a thickness equal to or greater than the projection.

'P'

↕ Not less than 'P'

◄——► At least twice the wall thickness

STRIP FOUNDATIONS

In the latter part of the 19th century, it was common to find the external walls of houses built directly on the ground. Legislation towards the end of the 19th century required concrete foundations under the walls. Then, as now, the depth of the foundation would depend on local conditions.

Nowadays, the design of foundations is controlled by national Building Regulations. Strip foundations, the most common form, can either be 'traditional' or trench-fill (see page 4). They are usually 500- to 700mm wide and as deep as necessary for the type of ground. In clays, for example they are usually at least 1m deep to avoid problems of ground movement caused by seasonal change in moisture content. In very dry conditions, for example, clays will shrink slightly as the clay loses water. In very wet conditions, the clay will swell. In weaker ground, the foundation has to be wider than 700mm to spread the building load over an adequate area of ground.

Requirements for strip foundations

The Building Regulations set out a number of requirements for strip foundations. Their width is determined by a table in the Regulations which takes into account building load and the nature of the ground. Their depth depends on site conditions and the nature of the soil – at least 1000mm is normal in clays. The Regulations also contain requirements regarding thickness of the concrete, position of the wall relative to the foundation, minimum depths near drains and so on.

In some situations, strip foundations are unsuitable, or are not cost effective. These include:

- Where large trees are present in clay soils (see page 5).
- Where trees in clay soils have recently been removed.
- In very weak or unstable soils.
- Where strip foundations would have to be very deep to reach firm ground.
- Where subsidence is likely (ie in mining areas).

In these situations, other foundation solutions, in other words piles or rafts, need to be considered.

Strip foundations

The width of a strip foundation depends on the building load and the nature of the ground. The depth should be sufficient to avoid the risk of frost attack or ground movement caused by seasonal change. In firm clay soils a width of 500–700mm and a depth of 1000mm is typical.

Both photos above show strip foundations. The one on the left is a 'traditional' strip, and the one on the right is trench-fill. The choice of either type largely depends on a number of practical considerations. Where the ground is not level the foundation can be stepped to avoid deep excavation (immediate left).

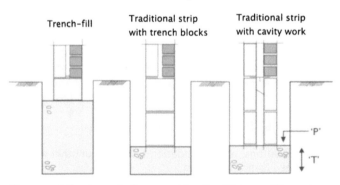

Trench-fill | Traditional strip with trench blocks | Traditional strip with cavity work

To prevent the concrete from cracking the thickness (T) must not be less than the projection (P) and never less than 150mm.

Trees and clay soils

Clay

Large trees will extract considerable amounts of moisture from the ground. In long, hot summers this may reduce the moisture content of clay soils. Shrinkage can lead to foundation settlement. Removing large trees can cause the reverse problem. Clay soils will slowly take up moisture formerly used by the tree. Ground heave can force part, or all, of the house upwards. Both problems will cause cracking.

Imagine a line from Hull to Exeter. A large proportion of the ground to the east of this line contains shrinkable clays. There are some belts of clay to the north of the line as well.

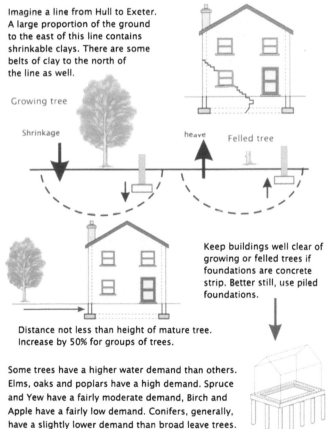

Growing tree

Shrinkage

heave

Felled tree

Keep buildings well clear of growing or felled trees if foundations are concrete strip. Better still, use piled foundations.

Distance not less than height of mature tree. Increase by 50% for groups of trees.

Some trees have a higher water demand than others. Elms, oaks and poplars have a high demand. Spruce and Yew have a fairly moderate demand, Birch and Apple have a fairly low demand. Conifers, generally, have a slightly lower demand than broad leave trees.

PILING

Piles are sometimes made from steel or timber although in most housing work piles are made from insitu or pre-cast reinforced concrete. Piles transmit the building loads deep into the ground. In clays, replacement piles are the norm; here the ground is augered or bored-out and replaced with reinforced insitu concrete. In sands and gravels, displacement piles are more likely to be found; in other words, the piles are driven into the ground, rather like driving a nail into a block of wood. Displacement piles are often made from reinforced pre-cast concrete. They can also be formed in insitu concrete where a steel cylinder is driven into the ground and removed as the concrete is poured down its centre. In housing, where the roof and floor loads are carried by brickwork rather than columns, a concrete beam sitting on top of the piles distributes the load from the brickwork into the piles themselves.

Some 20 or 30 years ago, piling was comparatively rare for housing (other than medium and high-rise flats). Since then, several factors have led to an increase in the use of piled foundations. These include:

- The increased pressure to re-develop 'brownfield' sites, where strip foundations may not always be appropriate.
- Increased costs of 'carting away' and tipping surplus excavation from foundation trenches (particularly in cities).
- The development and easy availability of smaller piling rigs and piling systems which are, nowadays, cost effective for house foundations.
- Greater understanding of piling in general (partly through better building education).

Factors affecting choice

There are literally dozens of piling companies in the UK, each offering a number of different piling systems. In many cases, more than one piling system will suit a particular set of circumstances. However, when choosing a piling system there are four main criteria to consider:

- building load
- the nature of the ground (ie the subsoil)
- local environmental or physical constraints (noise restrictions, height restrictions)
- cost.

Displacement piles

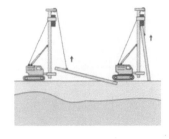

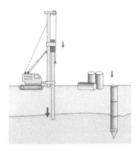

There are various types of displacement piles. The ones shown here are pre-cast concrete, either in long lengths or in shorter, cylindrical shells. They are driven into the ground until they meet a specified resistance. They are generally suitable for sands, gravels and some fills. They are not normally used in clay soils because of problems of whiplash in the piles as they are driven into the ground.

A reinforcement cage is positioned over, and tied to, the piles. A concrete beam is then formed over the piles. This takes the load from the walls and transfers it onto the piles, and then into the ground.

Replacement piles

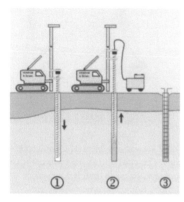

① ② ③

The augered pile is the most common form of replacement pile. These are commonly used in clay soils where displacement piles are generally unsuitable (see previous page on piling). The auger drills into the ground until it reaches the correct depth (1). It is then slowly removed, bringing to the surface the excavated material – at the same time concrete is pumped down its hollow stem (2). A steel reinforcement cage is then dropped into the pile (3).

Raft foundations

Rafts spread the building load over the whole ground floor area. In the 1950s and early 1960s they were often used to provide foundations for prefabricated housing. They are not very common in modern construction.

Flat slab rafts (right) offer a number of advantages over strip foundations: no trenching is required, they are simple and quick to build, there is less interference with subsoil water movement, and there are no risks to people working in trenches. They are generally suitable for good soils of consistent bearing capacity. However, they are expensive to design and construct. On poor ground, rafts must be stiff enough to prevent excessive differential settlement. This usually requires perimeter and internal ground beams (bottom right).

Reinforced slab
Hardcore bed

RAFTS

In the 1940s and 1950s raft foundations were fairly common, particularly beneath the thousands of prefabricated pre-cast concrete or steel buildings erected during the years following the Second World War. Most of these houses were built on good quality farm land where the soil was generally of modest to high-bearing capacity. Rafts (or foundation slabs as they were sometimes called) were often used because they were relatively cheap, easy to construct and did not require extensive excavation (trenches were often dug by hand). In 1965 national Building Regulations were introduced for the first time (London still had its own building controls), but these did not contain any 'deemed to satisfy' provisions for raft foundations (as they did for strip foundations). Consequently, each had to be engineer-designed. As a result, they quickly fell out of favour. In modern construction rafts tend to be used:

- Where the soil has low loadbearing capacity and varying compressibility. This might include, loose sand, soft clays, fill, and alluvial soils (soils comprising particles suspended in water and deposited over a floodplain or river bed).
- Where pad or strip foundations would cover more than 50% of the ground area below the building.
- Where differential movements are expected.
- Where subsidence due to mining is a possibility.

FOUNDATION FAILURE

This section examines the main causes of foundation failure. Extensive failure will usually result in building movement and cracking. However, cracking in buildings occurs for a variety of reasons and many of them have nothing to do with the foundations. (See the chapters on Cracking and External Loadbearing Walls for more information.)

Seasonal movement

Shallow foundations in clay soils are likely to crack as the ground rises and falls due to seasonal moisture change. If vegetation is present, substantial shrinkage of the ground may occur in long, hot summers as the tree slowly drains the water from the ground. If trees have been removed, the ground will swell and properties built with shallow foundations will crack. Cracked and leaking drains can cause similar problems. Cracking and movement caused by shrinkable clays account for most foundation-related insurance claims each year.

Frost heave

Some fine sandy soils, silts and chalk are likely to be subject to frost attack in cold weather. As the layers of water within the soil freeze, the associated expansion causes the ground to heave. It is unlikely to occur in occupied houses due to the heat loss from the floor but may occur in properties under construction, garages and garden walls. Insulated floors, which are becoming increasingly common, may well exacerbate the problem if foundations are too shallow.

Geological faults and mining subsidence

Differential movement of ground strata caused by earth tremors will usually lead to failures of traditional foundations. This is relatively rare in the UK but obviously can have devastating effects in those

areas of the world where earthquakes are more common.

A more common cause of ground movement in the UK is subsidence. There are literally thousands of old mineworkings in the UK – most of them dating from the 19th century. Unfortunately, in many cases, there is no record of their location. At the site investigation stage, local knowledge is invaluable.

If mines are known to lie under the site, the solution can range from filling them in with stone or concrete, using rigid foundations or using a building system that allows the building to cope with the movement. The picture shows part of Droitwich High Street. Underground salt extraction accounts for this problem.

Variable ground conditions

Where different strata exist in close proximity uneven settlement can lead to cracking. This can occur when building partly on filled areas or where natural changes in the ground occur, such as a small outcrop of rock. Even if this is not noticed at the trial pit stage, it should be identified as the foundations are excavated.

Differential movement

Similar problems of cracking can occur if houses have foundations of uneven depth and an example of this is houses with partial basements (see bottom picture on page 10). There is a danger that the shallow foundation may settle at a greater rate than the deep foundation and this differential movement will cause cracking. Similarly, when extending houses it is wise to ensure that the foundations to the new extension are at the same depth as the original property to prevent future problems of differential settlement. The porch extension (above right) was added several years after the original construction. It has quite shallow foundations.

The garage on the right was built on a shallow strip foundation. The cracking, which has occurred in both side walls, appeared within a few months of completion. However, that was nearly 40 years ago – it has not progressed since.

Chemical attack

Some soils, particularly clays containing a high proportion of sulfates, can attack the foundation concrete. In wet conditions, the sulfates combine with a by-product of the cement to produce a material called tri-calcium sulfo-aluminate. The crystals of this compound expand as they form and gradually cause pieces of concrete to break away. It can be prevented (but not cured) by the use of sulfate-resisting cement and by protecting the external faces of the concrete with a bituminous compound.

Building alteration

When buildings are altered, a situation can arise where the foundations are subject to increased loading. In some circumstances, this

can cause cracking of the foundation. In housing, it is most likely to occur when large panels of brickwork are removed to install patio doors. The columns of brickwork either side of the opening have to carry increased load and the extra force can crack the foundation concrete.

In some houses similar problems are caused by inadequate foundation design. The central pier of brickwork in this house has settled because the foundation below cannot carry the load.

Soil creep

On sloping ground with clay soils there is a danger that the upper layers of clay can slowly move downhill. It can often be recognised at the site investigation stage by the terraced appearance of the ground. Soil creep can occur in gradients as shallow as 1:10 and buildings with shallow foundations can obviously be damaged. Specialist advice is usually necessary when building on soils of this type and solutions can include retaining walls and piled foundations.

Tree roots

The problems of trees in clay soils have been covered in an earlier part of the chapter but, in some instances, there is a danger of roots themselves cracking foundations. It is particularly common in garden walls. The roots can also break drains, leading to saturation of the surrounding soil and possible soil heave or weakening of the soil.

Building on existing sites

If old foundations are not removed they can cause problems of differential settlement. This should, of course, be spotted at the site investigation stage.

Some existing sites may be contaminated by toxic substances. The ground can be 'capped off' with an inert layer, such as clean earth or stone, to isolate these substances if it is impractical to excavate and remove the contaminated ground. Foundations that rest on contaminated ground will need special protection and traditional gardens may need replacing with 'hard' landscape, such as paving, to prevent disturbance of the chemicals.

Poor workmanship

A variety of problems can occur on site which will lead to future foundation failure and these are usually a result of poor workmanship and inadequate supervision. Typical problems include:

- incorrect mixing of concrete
- use of contaminated aggregates or old cement
- inadequate cleaning out of trenches before concrete is poured
- concreting in freezing weather
- adding excess water to help the concrete flow round the trench.

CHECKLISTS

Most chapters in this handbook include a simple checklist to provide systematic guidance to anyone carrying out a building inspection. However, a checklist is not much use for examining the foundations. For a start, you can't inspect them. In addition, the manifestation of serious foundation failure is usually self-evident, even if the underlying cause is not. If you see signs of recent movement or evidence of substantial cracking, specialist advice will always be necessary. The chapter on Cracking may help you to diagnose the problem but remember this book is not a substitute for professional advice.

EXTERNAL LOADBEARING WALLS

EARLY BRICKWORK

By the early 1700s brick buildings had become common in many parts of the UK. In towns and, to a lesser extent, in rural areas the timber frame tradition was slowly coming to an end. Timber (home-grown) was in short supply; there were concerns about the risk of fire, and the fashion for classical buildings did not suit the rustic, organic and irregular nature of timber framing.

These two buildings were built only 40 years apart; the one top left (1670s) is timber framed; the one bottom left is an example of the Queen Anne style (about 1710).

During the 1700s there were a number of improvements in brick making. Blended clays, better moulding techniques and more even firing gave greater consistency in brick shape and size. Fashion dictated brick colour: the reds and purples popular in the late 1600s gave way to softer brown colours in the 1730s. By 1800 the production of yellow London stocks provided a brick colour not that different from some natural stones.

The repeal of the brick tax in 1850 gave the brick industry a new impetus. Improved mixing and moulding machines, together with better firing techniques, allowed brick production to reach new heights. Bricks were now available in a range of colours, shapes and strengths that would have been unimaginable a 100 years earlier. Better quarrying techniques allowed extraction of the deeper clays, which produce very strong, dense bricks; vital for civil engineering works such as canals, viaducts sewers and bridges.

BRICK BONDING

To guarantee effective weather protection and provide adequate strength the walls of houses need to be at least 215mm thick – these are known as 1 brick thick walls. Houses over three storeys often had thicker walls, usually reducing in thickness at each floor level.

The brickwork itself was generally laid to a very high standard. Most houses were built in Flemish bond although Header bond was also

popular in the early to mid 1800s. The next page shows the construction of a wall in Flemish bond. Note that all solid wall bonds require the use of special bricks, known as Queen (& King) closers to complete the bond at window openings and corners. English bond (the strongest bond) was generally reserved for engineering works although its weaker sibling, English Garden Wall bond, was used extensively at the rear of properties or behind renders. An example of English Garden Wall bond can be seen later (see Openings in solid walls on page 20).

Despite its name, English Garden Wall bond is not uncommon in older housing where it is often found forming the rear walls of houses (sometimes hidden by render). Its popularity lay in its cost. It is easier to lay and level stretchers and there is less waste because oversized bricks (which might have to be rejected as headers) can be accommodated by 'stretching' the joints.

Flemish and English bond are by no means the only bonds used in the construction of solid walls. By varying either bond slightly others can be formed.

Flemish bond

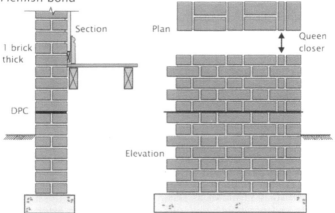

Section · Plan · Queen closer · 1 brick thick · DPC · Elevation

Section

1.5 brick thick

Nearly all brick houses built before the 1920s were built in Flemish bond - or at least the walls on show to the street were. In this bond a single course of brickwork contains alternating headers and stretchers. Most two and three-storey houses were built with walls 1 brick thick (see above example). Houses of more than three storeys usually had thicker walls. Flemish bond can easily be adapted to suit walls of 1.5 brick thick and 2 brick thick.

Header bond is fairly rare but it is an alternative to Flemish bond.

In Flemish bond it was not uncommon to find headers and stretchers in different colours.

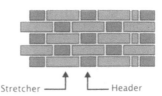

Stretcher ——— ⬆   ⬆ ——— Header

MORTAR

Mortar is a mixture of materials for jointing masonry units. It has numerous tasks. It sticks the bricks together to provide stability and solidity – while holding them apart to spread loads evenly. It compensates for irregularity between units when straight, level and plumb walling is laid. It also seals any gaps to resist wind or rain penetration. As well as fulfilling its gap-filling adhesive function, it should also have adequate durability and strength.

Material for mortar

Lime mortars were common until the 1930s and are still used today, particularly in conservation work. Limestone or chalk is burnt with coal (calcination) to form calcium cxide – Quicklime. During burning carbon dioxide is 'burnt off'. The burnt lime is known as lump lime. The next stage is known as slaking. In this process the quicklime is added to water where a vigorous chemical reaction (which produces great heat) converts the quicklime into slaked lime. Slaked lime is calcium hydroxide. The slaked lime is mixed with fine aggregates (nowadays sand) to form mortars, renders and plasters. As the material slowly sets it combines with carbon dioxide in the atmosphere to return to is original state – calcium carbonate. It can take many months for a lime plaster to fully set – hence the long delay before decorations common 50 years ago. This whole process is sometimes referred to as the triangle of lime. Slaked lime can be stored almost indefinitely.

Some limes have a hydraulic set (usually in addition to, not in place of, carbonation). This can be induced by adding pozzolans – which contain silica. There are a number of materials and industrial wastes, which can act as pozzolans. Another option is to use a lime that naturally contains silicas (usually a proportion of clay). Hydraulic sets are quicker and stronger than carbonation. Some of the very strong hydraulic limes are not dissimilar to cement (made of course, from chalk and clay). By the end of the Victorian period and during the early part of the 20th century, brick mortars were often based on hydraulic lime rather than the slower-setting, non-hydraulic or putty lime. A typical mix for external walls would be one part hydraulic lime and three parts sand (or other locally available aggregate). Cements of various types (but mostly made from a mixture of lime and clay, fired at high temperatures) were also available and were generally recommended where extra strength was required or for use below ground level.

POINTING

Nowadays, most brickwork, is jointed as work proceeds. This was also the norm for much of the terraced industrial housing built during the second half of the 19th century. Pointing, a much more expensive operation, is the term used to describe existing or new joints that have been raked out and filled with fresh, often coloured, mortar. The pointing mortar mix must be slightly weaker than the jointing mortar. If it is stronger the outer face of the bricks, immediately above and below the pointing, will carry excess load. This can result in the edges of the bricks spalling.

In the Victorian period, joints were usually finished flush or slightly recessed. Where very good quality bricks were used the joints were often only 8mm, or even less. This, together with the use of brick dust in the mortar, meant that the mortar had very little affect on a building's appearance. Compare this with the Georgian townhouse shown inset. This wall (in uncommon Header bond) has recently been repointed in an unforgiving and unsympathetic cement mortar with proud joints.

Working-class housing was usually pointed in a lime mortar, which included local industrial waste products as fine aggregate. Ash was probably the most common.

Tuck pointing was usually reserved for the best quality work. Tuck pointing is basically in two parts, a bedding mortar often containing

aggregates to match the colour of the bricks or stonework, and a thin ribbon of lime pointing to finish the joint. From a distance a wall that is tuck pointed appears to be finely jointed.

OPENINGS IN SOLID WALLS

Where openings occur for doors or windows support is required at the top, or head, of the opening. A window frame is not designed to carry any wall loading. The load that has to be carried is, in fact, quite modest due to the bonding of the brickwork. The loads can be supported by arches or by stone lintels. A common solution is shown in the photograph below. Here a segmental arch (with a timber lintel behind) caries the loads over the opening. Arch construction can be quite complex and there are many styles and techniques. In housing, arches are normally segmental in modest and working-class housing and semi-circular or even flat in better-quality housing. Arches can be constructed in three main ways. Rough arches are formed by using ordinary bricks and tapered mortar joints. Axed arches are formed by cutting the brick to a taper on site (no taper in the joints). Gauged arches are formed by using special soft clay bricks that can be rubbed to a very accurate taper on site. They are laid with very fine joints. A number of examples are shown on the next two pages.

Arches and openings

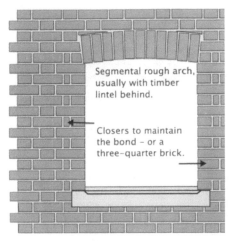

Segmental rough arch, usually with timber lintel behind.

Closers to maintain the bond – or a three-quarter brick.

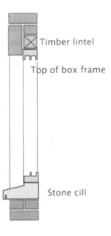

Timber lintel

Top of box frame

Stone cill

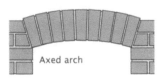

A rough arch formed in half-brick rings.

Axed arch

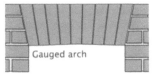

Gauged arch

If snapped headers are used to form the arch the mortar joints are not quite as wide at the top. A typical segmental arch rises about 25mm for each 300mm of span.

If the bricks are tapered slightly the joint thickness can be maintained. Where the bricks have been roughly cut or shaped on site they are known as axed arches.

Where soft bricks have been 'rubbed' to shape and where very fine joints have been used they are referred to as gauged arches. Axed and gauged arches can be curved or flat.

A rare but very fine example of a gauged ogee-shaped arch.

A gauged semi-circular arch over a door opening. This example is from the late 18th century. Semi-circular openings over windows are fairly rare as they do not suit the nature of sash windows. Notice the very fine joints in the brickwork itself.

The photograph on the left shows a simple, rough segmental brick arch. Below (left) is a late Victorian brick-ring arch in the pointed Gothic style. The photograph directly below shows a more decorative segmental arch, but technically still a rough arch due to its tapered mortar joints.

Stone lintels could also be used at the head of an opening. They could be flat or segmental. This example has been cut from a single stone and includes a 'fake' keystone for decorative detail. Note that the stone cill also has two consoles below it. The indented brickwork under the eaves adds further decoration to the facade.

STONEWORK

Stone has been used for building in Britain since at least the first century BC, and earlier, in surviving stone structures such as Stonehenge. The Romans used both stone and brick for their buildings, although the use of these materials declined during the so-called Dark Ages after the Romans left. The Saxons reused Roman brickwork (at Brixworth, Northants) and later the Vikings used stone for their important buildings, especially churches, such as Escombe, Barton on Humber and Earls Barton in the North and Bradford on Avon in the South. The Normans used stone fairly extensively; many churches and cathedrals still have Norman elements in their buildings, some fine examples being the church at Kilpeck, and Durham, Gloucester, Canterbury and St David's cathedrals. There is not much evidence, however, to suggest that stone was used extensively for domestic buildings before the 14th century.

The following centuries saw an increased use of stone for dwellings, as various sectors of society became wealthier, especially those involved in the wool trade. Examples of wealthy merchants' houses can still be seen, such as the fine examples in Chipping Camden, in the Cotswolds.

In the rebuilding of London after the Great Fire and during the Georgian period, stone was much prized. Provincial cities such as Bath and Edinburgh boast many fine Georgian buildings, most of them built in the classical style.

By the Victorian period, brick was a common building material but stone was still used for the houses of the wealthy, other prestigious buildings, and for decorative work on the facades of more humble housing. The use of stone for buildings has always suggested power, wealth and authority, and some of the finest stone buildings of the Victorian period are the banks, where the fine stone work was designed to suggest a stable and trustworthy business.

In upland areas (the north and west), stone was often the obvious choice for building because it was readily available (and prior to the railways these were often areas where bricks were expensive). In these areas, housing for most people was built in the vernacular tradition, not adhering to any 'rules' of classical design.

There are three groups of stone: igneous, sedimentary and

metamorphic. The sedimentary group, which includes limestone and sandstone, accounts for most of the stone used for building in the UK.

Rubble walling

Rubble walling is found in a variety of styles. At its cheapest it comprises rough stonework, built as two outer leaves and bound together with copious amounts of lime mortar.

More expensive work comprised squared rubble possibly set against a brick backing.

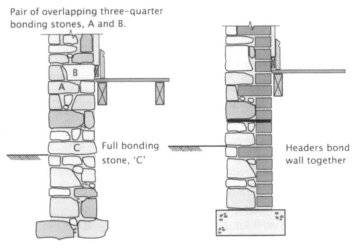

Pair of overlapping three-quarter bonding stones, A and B.

B
A

Full bonding stone, 'C'

Headers bond wall together

Simple rubble wall   Rubble wall with brick backing

In most situations, a stone wall has to be thicker than a brick one. So, whereas a 1 brick thick wall (215mm or so) might be fine for a two or three-storey house, a stone wall is likely to be 375mm or even more.

Rubble walling is, very much, a local tradition. Styles, terminology and materials are unique to geographical location. However, there are some basic walling styles that most surveyors should recognise, see opposite page.

The picture above shows a rubble wall from Exmoor. It is laid in a weak lime mortar. The picture on the right shows a section through a rubble wall; copious amounts of mortar are required to fill all the gaps. The random rubble wall below (left) is an example from the South-east. The one on the right is a Victorian Gothic example – technically a rubble wall but here every stone has been 'designed'.

The pictures below show two examples of squared rubble from central Wales. The picture on the left shows work that has been partly brought to courses – this is sometimes called snecked rubble. The example on the right shows rubble that has been properly coursed.

Where window openings occur in rubble walling the support at the head is generally of three types: stone arches, stone lintels, and brick arches. The photograph on the left shows an example of a rough segmental arch.

Finally, note that most rubble walls were pointed flush or slightly recessed. The ribbon pointing so often seen nowadays is not traditional, neither is it particularly durable.

Dimensioned stone

Stone that is dressed and/or finely cut is often referred to as dimensioned stone. It is sometimes referred to as freestone. This means it can be worked (cut, shaped and smoothed) with a chisel and a saw in any direction. It has a fine grain and is free from obvious laminations and pronounced bedding planes. In the 18th century, whole cities were built (some rebuilt) in stone. It was not cost effective to build the whole of the wall in freestone and a backing material of rubble or brickwork can nearly always be found. In some houses only the front elevation would be built in freestone, the sides and back being constructed of rubble or brick. To bond the two halves of the wall together, 'through' or bonding stones were used. The backing material would largely depend on the locality. In Bath and Bristol for example, there are several limestone quarries with cheap rubble stone available. In London, near to clay beds, brickwork was much cheaper. In some forms of construction, the facing stone is tied back to the brickwork/rubble with iron ties.

These were quick to rust and could soon result in bulges appearing in the facework.

Many brick or rubble houses contain some freestone ornament or decoration. Perhaps the most common elements are window surrounds, string courses and cornices. The example on the right is late Victorian.

Nearly all freestone work is limestone or sandstone. Some of the most 'famous' limestones are Ancaster, Clipsham, Bath stone, and Portland stone. Portland stone was used extensively in London in the 17th and 18th centuries. It is off-white in colour, with a fine even texture. It is also one of the most durable limestones available – well suited to the pollution in a big city. Important sandstones include Craigleith (used in Edinburgh), Wealden and the red sandstones from the Midlands.

Sedimentary stones are normally bedded in the same way they were naturally formed. In other words, the 'layers' within the stone are horizontal. There are exceptions to this rule. If the cornice (see drawings), for example, was naturally bedded it would soon deteriorate. In this example the cornice and parapet blocking are edge bedded. Stone should never be face bedded, in other words with the layers vertical but parallel to the stone face, as it will soon delaminate.

Where the freestone is laid with very fine joints, almost invisible from more than a few feet away, the work is known as ashlar. In some parts of the country the stones were cut with a taper to make the joints easier to form. Wedges made from bits of timber or even oyster shells were often pushed into the back to provide stability as the mortar set. These buildings were built with lime mortar, which hardened very slowly. Hydraulic limes were not unknown but they were less common and more expensive. In addition, they often set too quickly, resulting in high waste on site.

Ashlar

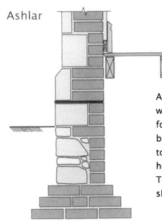

A typical late Victorian or Edwardian ashlar wall with a brick backing and a stepped footing. The two halves of the wall (ie the brick and the stone) must be well bonded to ensure stability and durability. In some houses the ashlar is secured by iron ties. These are less satisfactory than the method shown here.

Sedimentary stones are 'laid down' over thousands of years. Where possible the stones should be bedded in the same plane as they were cut from the quarry. However, some exposed stones are more likely to weather if bedded this way.

The picture below shows a limestone quarry. This same quarry (Doulting) provided the stone for Wells Cathedral in the 13th century.

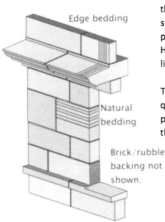

Edge bedding

Natural bedding

Brick/rubble backing not shown.

String course

Section and elevation

In humbler brick houses freestone window surrounds, cornices and string courses often add a bit of decoration to what, otherwise, might be a dull facade.

CAVITY WALLS

In the latter part of the 19th century, a number of houses were built with cavity walls. It was not, however, until the 1920s that this became the accepted form of construction. Cavity walls are cheaper than their solid wall counterparts. In addition, they offer improved thermal insulation and better weather protection. Most walls comprised two half-brick leaves with a 50mm cavity. The two halves of the wall were tied at regular intervals with steel or wrought iron wall ties. The external

Wall Ties

Wall ties from a 1925 trade catalogue. The upper ties were available tarred or galvanised. The lower ties (wire) were galvanised only.

leaf of brickwork was laid in facing bricks, the internal leaf in commons. A 1929 ironmonger's catalogue lists two types of tie (above).

Bricklaying mortars were usually based on hydraulic lime although cement (perhaps with the addition of some non-hydraulic lime to improve its workability) was becoming increasingly common.

To prevent moisture reaching the internal leaf damp proof courses (DPC) were included in the construction. The drawings on the next page show typical details. Window heads were formed in a number of ways: segmental arches were still common although they were slowly replaced by concrete and steel lintels, which were cheaper and quicker to fix.

The volume house-builders built millions of houses, public and private, during the 1920s and 1930s. Unlike Victorian houses, these new houses no longer boasted any regional characteristics. Magazines were full of adverts for bricks, tiles, windows and other products.

RUSTIC FACING BRICKS

LONDON BRICK
COMPANY & FORDERS LTD
AFRICA HOUSE, KINGSWAY, LONDON, W.C.1

Early cavity walls

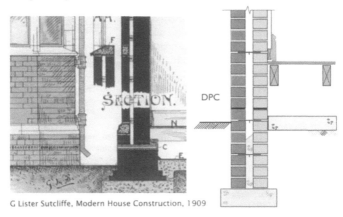

G Lister Sutcliffe, Modern House Construction, 1909

Some very early cavity walls (1890s) were built with a 1 brick thick internal wall (see drawing above). However, by the 1920s two half-brick leaves were the norm. Both leaves were usually laid in stretcher bond although, in some cases, the external leaf was laid in Flemish bond (with snapped headers – done to emulate more expensive 1 brick work).

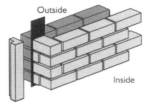

The vertical DPC prevents water form crossing into the internal leaf. Many early cavity walls (1920s) did not include vertical DPCs.

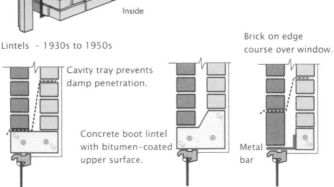

Lintels – 1930s to 1950s

Brick on edge course over window.

Cavity tray prevents damp penetration.

Concrete boot lintel with bitumen-coated upper surface.

Metal bar

During the 1950s and 1960s construction did not change significantly. Blockwork replaced brickwork for the internal leaf of the cavity and steel lintels slowly replaced concrete ones.

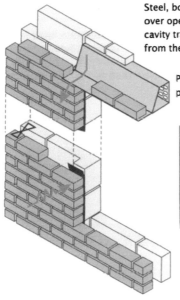

Steel, box-section lintel carries loads over opening. The sloping face forms a cavity tray to direct any water away from the internal leaf.

Perforations on the back of the lintel provide a key for the plaster.

The frame is fixed to the external leaf of brickwork with cramps.

MODERN CAVITY WALLS

In modern cavity work, the most significant development has been the addition of insulation. A modern cavity wall has a 'U' value (a measure of the wall's ability to transmit heat) some 5 or 6 times better than its 1920s counterpart. This is, in part, due to the development of insulating materials designed to fit inside the cavity, and the use of aerated concrete blocks for the inner leaf.

Blockwork

Blockwork has become very popular in the past 60 years or so because of its cost advantages over brickwork. There have been several aggregates used in their manufacture, such as crushed gravel, pulverised fuel ash, blast furnace clinker, gas coke breeze and pumice. Changes in industrial processes have meant that some of these aggregates are no longer available. Dense blocks, suitable for party walls, internal loadbearing walls and foundation work, are made from cement, sand and crushed gravel. Until the 1980s, they were also used for the internal leaf of cavity walls. Lightweight blocks are made from cement and various lightweight aggregates and are generally used for the internal skins of cavity walls.

In modern construction, aerated concrete blocks have become very popular They are made from cement, lime, sand, pulverised fuel ash and aluminium powder. Once these materials are mixed with hot water the aluminium powder reacts with the lime to form millions of tiny pockets of hydrogen.

The photographs below show dense blocks on the left and lightweight, aerated blocks on the right.

Modern mortars

Modern mortars are made from cement and sand. Hydrated lime (ie bagged lime) is often introduced into the mix to give it a more plastic feel and to make it more 'workable'. Lime also improves the mortar's ability to cope with thermal and moisture movement. By varying the proportions of the cement the strength of the mortar can be increased or decreased as required. As a general rule a weak mortar mix will suffer minimum shrinkage when it sets and is more able to withstand the long-term rigours of thermal and moisture movement. In addition, if minor cracking does occur in the wall, it is more likely to occur at the joints which can easily be raked out and repointed. However, the mortar must not be too weak or it will become too porous and may crush under high compression forces.

For most brickwork, a mix of 1:1:6 (cement:lime:sand) will provide a durable mortar. Various chemicals can be used in place of the lime to provide a mortar with similar properties.

In recent years, the use of pre-mixed mortars has become common.

These are delivered to the site in sealed containers, ready for use. They usually contain a retarder so they remain usable for 36–48 hours or so. At the end of this period, they develop their strength in the same way as normal mortars.

Jointing and pointing

The face of the joint (pointing is almost unknown nowadays) may be finished in a number of ways – the three most common are shown below. Tooled joints (where the mortar is pressed against the brickwork) offer the best weather protection because the tooling smoothes and compresses the joint.

Bucket handle   Flush   Recessed

Wall ties

Since the 1960s, wall ties have been made from a number of materials, including galvanised steel, stainless steel and plastic. Today, most ties are made from stainless steel. The spacing of the ties is typically 900mm apart horizontally and 450mm vertically. Wider cavities (they are now common up to 100mm wide) may need additional ties. Additional ties are also required around window and door openings.

Insulation

There are three common options that will meet current requirements all of which require lightweight or aerated blocks in the inner leaf. These are:

- a clear cavity with an insulated dry lining
- insulation boards that partially fill the cavity
- insulation batts that fill the cavity.

Solid walls are also acceptable: thick aerated blocks can easily achieve required insulation levels although additional external weather protection must be provided.

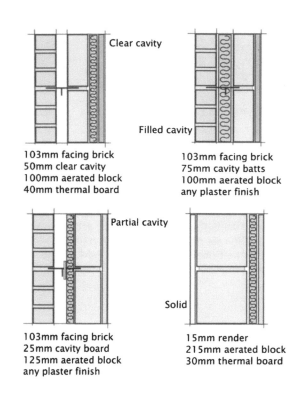

Clear cavity

103mm facing brick
50mm clear cavity
100mm aerated block
40mm thermal board

Filled cavity

103mm facing brick
75mm cavity batts
100mm aerated block
any plaster finish

Partial cavity

103mm facing brick
25mm cavity board
125mm aerated block
any plaster finish

Solid

15mm render
215mm aerated block
30mm thermal board

Insulation can be provided in a number of ways. Two are shown here: the one on the left is a foil faced 'bubble wrap', the one on the right are polystyrene cavity boards.

Loads over windows and doors are normally carried by steel lintels (with integral insulation). The brickwork over openings is sometimes 'on edge' to provide a decorative feature. Notice the weep holes – to allow the escape of any water finding its way into the cavity.

Modern DPCs are usually made from polythene.

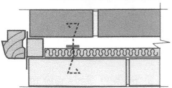

Vertical DPCs can still be used to separate the leaves. This plastic, extruded strip is another option.

Structural stability

In domestic construction, the stability of an external wall depends on a number of criteria. These include its height and length, correct wall tie provision, the 'bonding in' of partition walls and party walls, and adequate restraint from floors and roof structure. These items are considered in more detail in the next section.

DEFECTS IN WALLS

NB: problems of damp penetration and condensation can be found in the chapter on Dampness. Cracking caused by foundation failure is dealt with in the Foundations chapter (see also Cracking chapter).

PROBLEMS OF SATURATION

Chemical attack

Efflorescence is usually a temporary occurrence on new brickwork and in most cases, is harmless. It can be recognised by its white powdery appearance. It is caused by the small amount of alkaline salts present in the clay brick which dissolve after rainfall and are brought in solution to the face of the brick. After several months of rainfall they usually disappear.

Sulfate attack of the mortar joints is a much more serious condition and is likely to occur in areas where certain types of brickwork are very exposed or in direct contact with the ground. It is a chemical reaction

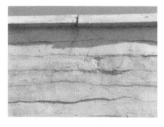

caused by the sulfates in the bricks dissolving in rain or ground water and washing into the brick joints, where the solution reacts with the alumina in the cement (present in the mortar) and forms a compound that causes rapid expansion. It can be easily recognised by pronounced horizontal cracking in the bed joints.

It is a common problem, particularly in parapet walls and unlined chimneys. In the latter, the problem can be exacerbated by condensation of exhaust gases adding both water and sulfates. The condensation occurs on the cold side of the chimney (usually the north side) and this, together with uneven drying due to orientation, can cause the chimney to lean at alarming angles.

Lime staining occurs when calcium hydroxide is deposited on the face of brickwork. It quickly absorbs carbon dioxide from the air and becomes calcium carbonate. It is virtually insoluble and can only be removed by expert and costly treatment. The calcium hydroxide can come from three main sources:

- calcium hydroxide caused by the hydration of Portland cement
- from hydrated lime added to mortar
- from brickwork in contact with wet concrete.

Spalling and frost attack

These problems often occur in low-quality, porous bricks and are caused by saturation of the bricks. In the case of spalling, the problem is caused by the continual contraction and expansion of the brick face as a result of moisture movement. Eventually the face of

the brick breaks away. It is very common in exposed situations. With frost attack (see photo) the face of the brickwork can be forced away as absorbed water freezes and turns to ice. The face of a poor-quality brick can also break away due to the expansion of chemical impurities it may contain. Notice the mortar is largely intact.

Wall tie failure

Some home owners and local authorities are experiencing problems of wall tie failure. Although most steel wall ties are galvanised (coated with zinc) the zinc protection breaks down over the years due to natural electrolytic action. This occurs where two metals are in direct contact, ie the zinc and the steel. Those ties with a thin layer of galvanising are most at risk. Without the protective galvanising the tie is free to rust. The rate at which the tie fails depends on a number of factors, which include the nature of the mortar and the exposure of

the building. As rust occurs the wall tie expands forming horizontal cracks in the brick joints. This is usually easy to recognise as the cracks occur every sixth course of bricks, the usual spacing of ties. In some cases, the centre of the tie is the first part to rust and this can lead to severe structural problems as the wall effectively becomes two half brick skins. Several systems of inserting stainless steel ties or bolts have been developed to deal with this problem.

Material failure

Bricks and blocks

New clay brickwork is likely to expand for several weeks after manufacture, with most expansion occurring in the first few days as it absorbs moisture from the atmosphere. If the bricks are used before this expansion has taken place the bricks will continue to expand when in position and the wall may crack.

Calcium silicate bricks suffer from the opposite problem, ie they tend to shrink in size for the first few weeks following manufacture. The photograph on the right shows cracking in a run of wall without an adequate number of movement joints. Similarly, concrete blocks that are new or have been stored uncovered in the rain are likely to

continue to shrink once laid. It is wise to wait as long as possible before plastering. A weak mortar will, to some extent, minimise the effects of shrinkage but some lightweight blocks will crack as they dry out.

Mortar and pointing

The nature and finish of the mortar joints are probably more significant than the nature of the bricks themselves in terms of a wall's durability. If the mortar is too weak, it will be porous and subject to frost attack. If it is too strong, it will be unable to accommodate the thermal and moisture movement it will encounter throughout its life. Strong mortars are also prone to drying shrinkage;

in some cases, shrinkage can break the bond with the bricks around it. The garden wall shown (top left) was repointed a few years ago in a very strong mortar mix (something like 2 parts sand to 1 part cement). Within 12 months the pointing had failed. Mortar finishes that stand proud are at risk from frost attack (besides being very expensive). Recessed joints (below left) should be avoided, particularly if the cavity is filled with insulation. The top edge of the brick can stay wet for long periods; there is, therefore, an increased risk of damp penetration and more likelihood of algae growth on the brick edges.

The bricks in good-quality buildings from the first half of the 20th century were often joined in a cheap mortar and then pointed in a more expensive one. If the pointing mortar is stronger than the jointing mortar the loading on the bricks is uneven and the edges (where the load is concentrated) are likely to spall. Pointing can also fail because it is not properly tooled or where it is of insufficient depth.

Structural movement

The majority of structural problems can occur as a result of movement in the ground and this has been covered in the previous chapter. There are, however, some problems that can occur due to poor design or bad workmanship above ground level. These problems usually manifest themselves as cracking, buckling or bulging, and distortion.

Thermal movement

Over the course of an average year, an external wall will suffer extremes of heat and cold. As the wall heats up, it will expand and as it cools it will shrink. Similar expansion and contraction occurs due to changes in moisture content. At the design stage this movement must be allowed for otherwise cracking will occur. In long terraces of houses, particularly those built with a strong cement mortar, cracking caused by thermal and moisture movement is a common problem.

Terraced housing will expand due to increases in moisture content

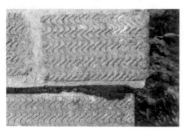

and temperature. You will often find examples where the superstructure walling, at the end of the terrace, has oversailed the DPC as a result of this expansion.

When environmental conditions change and the terrace starts to contract, cracks will often appear at the planes of weakness, usually around windows and doors. This is due to the fact that the bricks and mortar are low in tensile strength and cannot pull the terrace back into shape. Older properties, built using lime mortar, are less prone to this form of cracking as the weak mortar can accommodate minor movement.

While probably not structurally dangerous, the cracks do allow moisture to penetrate the wall. To prevent expansion and contraction from causing damage it is usual to provide movement joints in large panels of brickwork.

Inadequate thickness and bond

In some situations (wall-tie failure mentioned earlier is one of them), a solid wall or cavity wall will fail across its width. This can manifest itself as buckling and bulging, sometimes on the inside face and sometimes on the outside face. Two fairly common examples are shown below (another is shown on page 45). The first example shows a rubble wall with inadequate bonding stones. The wall is, in effect, acting as two separate leaves. Built like this, it is less able to withstand vertical and horizontal loading. The second example shows a similar type of failure; this time in a brick wall. In some old examples of English Garden Wall bond the courses of headers are too far apart. Where buildings are rendered you will sometimes find the headers are almost non-existent.

It is quite rare, but not unknown, to find walls where the brick

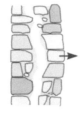

Rubble wall poorly bonded

'Separation' is sometimes most obvious around windows.

A wall built in English Garden Wall bond may have insufficient headers.

headers have been 'snapped'. This was sometimes done to give early cavity walls the appearance of solid walls. If a house, say, from the late 1920s or 1930s has face brickwork in Flemish bond think carefully before assuming it is a solid wall.

Some Georgian and early Victorian walls contain a large timber beam (called a bressumer), usually at first floor level. If this rots, it can affect the stability of the wall. Where bonding timbers have been built-in to the wall (either to support panelling or to speed up the construction of the wall) similar problems can occur. Lime mortar

Timbers built-in to support paneling can cause similar problems

Timber bressumers, often the full thickness of the wall, can cause buckling of the brickwork if they rot.

41

was very slow to harden and the use of bonding timbers helped the speculative housing developers finish their buildings quickly.

Lack of lateral restraint

Walls are comparatively thin and without various kinds of restraint they would soon topple over. The flank wall of a house should be restrained by the front and back walls, the upper floor (depending on

the direction of the joists), the roof, and possibly any internal walls and chimney breasts. As buildings age and settle, the restraint that is lost most readily is that from the first floor joists. Buckling and bulging can occur where this restraint is lost. The traditional solution was to insert iron tie bars throughout the property and fix large washers (the crosses in the photograph) to each end. This prevented further movement. Similar solutions are still used today – but using stainless steel not wrought iron.

Lack of restraint from floor allows wall to move outwards. This is a particular problem where stairs are against the outside wall. Buildings with more than two storeys are even more at risk.

Thrust from roofs

Similarly, at roof level the structure must be well secured to the wall and the rafters must be restrained from moving outwards. This restraint is usually provided by the horizontal ceiling joists. Look at the right-hand window and the line of the wall behind the fascia board.

Excess loading

Failure of arches and lintels is not uncommon. Four examples are shown below. The first example (top left) shows failure of an arch; in this instance failure has been caused by deterioration in the mortar joints, which has allowed the arch to drop. The second example (top right) shows a pier of brickwork that has dropped due to failure of the timber lintel over the wide bay window opening.

Carbonation (bottom left) is a defect affecting reinforced concrete. As the concrete slowly absorbs carbon dioxide from the atmosphere, it loses its alkalinity and any embedded steel starts to rust (it takes 40 years or so to manifest itself). As the steel expands, the concrete cracks. There is no repair other than replacement of the lintel. The last example (bottom right) shows a stone lintel that has fractured. The chapter on Cracking examines lintel failure in more detail.

43

Other problems

There are a few other 'induced' problems that affect external walls. Two are shown here. The first is caused by vegetation – some plants can cause severe disruption in the joints within the space of one or two growing seasons. If this problem had been spotted promptly, failure would have been avoided. Similarly, prompt action to fix the leaking overflow would have prevented this extensive and unsightly staining.

Stonework

There are three groups of stone: igneous, sedimentary and metamorphic. The sedimentary group, which includes limestone and sandstone, accounts for most of the stone used for building in the UK. Some of the specific defects in sedimentary stones are shown opposite.

Many of the problems in stonework are caused by ill-conceived remedial work. Pointing is the most obvious example. The picture below shows a rubble wall which, at some point in the past, has been repointed in a strong cement mortar. The white 'joints' are paint!

Stone defects 1

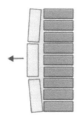

Ashlar work rarely forms the whole thickness of the wall. It is usually a thin facing (about 100mm or so) tied back to the rubble or brick wall with metal toes or with bonding stone If there are an inadequate number of ties or bonding stones, or if the ties rust the facing can part company with the backing.

Sedimentary stone is normally laid on its natural bed. If it is face bedded (as shown here) it is very vulnerable. The wetting and drying cycle, or freezing and thawing can induce stresses that force off the face.

Frost attack can quickly damage exposed soft stone, particularly if it is free stone. In this situation it should be protected with a lead cover flashing.

This carboniferous limestone has been repointed with ribbon pointing. This finish is historically inaccurate and potentially damaging to the stone.

Whatever the pointing style, cement mortars should be avoided with relatively soft free stone. Its dense, brittle and unforgiving nature usually causes irrevocable damage to the stone.

Stone defects 2

Limestones (and some sandstones) are at risk from sulfate attack. This occurs when acid rain (sulfuric acid) attacks the calcium carbonate in the limestone or in the binding matrix of sandstone – forming calcium sulfate. As this crystallises, it forms blisters.

This soft sandstone is next to a busy main road. The weathering may have been due to the wetting and drying cycle, freezing and thawing, or road salt.

Some limestones and sandstones can be affected by vegetation, birds and even bees. Masonry bees, for example, will bore into soft stone or mortar to nest.

These limestone columns show a 'ring' of damage about 500mm above pavement level. It is not caused by impact damage but by the crytallisation of salts carried up the columns by rising dampness.

In sandstones and limestones natural fissures and weak beds can occur. These should be identified and rejected by the mason but they are often found in speculative work.

CHECKLIST (EXCLUDING FLATS)

Element	Comments/problems
Construction	To help analysis try and identify property age, wall bond, material, thickness and whether the wall is cavity or solid. Render defects are dealt with in a separate chapter.
Structural stability and movement	Externally look for: • Cracking (see the chapter on Cracking). • Bulges and buckling (or signs of remedial work). • Walls out of plumb. • Thermal movement (probably in the form of vertical cracking). • Oversailing (most likely at DPC level or parapet). • Potential structural problems caused by saturation, eg wall tie failure and sulfate attack. Check inside the building for signs of cracking – the first places to check are junctions of: • Party wall and external wall. • Ceilings and external wall. • Partitions and external wall. If there are signs of buckling and bulging from the outside but the inside walls seem plumb, suspect bond problems within the wall itself. Check for building alterations or removals that may have affected restraint and stability, eg chimney breast removal and new openings in internal loadbearing walls. Check for signs of cracking or movement where large timbers (or steel beams) carrying point loads are built into the wall. Common examples include gutter beams, floor beams, trusses and purlins.

Element	Comments/problems
Openings	Look around openings for: • Dropped arches. • Cracks or movement in and around lintels. • Gaps between windows/doors and masonry (may indicate joinery movement or buckling of wall). • Cracks in sills.
Material failure – brick	Look for: • Spalling (most likely caused by frost attack). • Efflorescence. • Staining (chemicals within the bricks or run-off from other materials such as lead flashings). • Expansion of new brickwork. • Friable (crumbling and porous).
Material failure – stone	Look for: • Incorrect bedding. • Fissures in the stone. • Soiling (mostly from soot). • Staining (see bricks above). • Chemical attack.
Mortar	Is the mortar strength appropriate for the masonry? Is it 'complete' – in other words, is it bonded to the masonry; is any mortar missing? Is the pointing profile appropriate for the masonry? (Recessed pointing in brickwork and ribbon pointing in stone are the two worst culprits.)
Weather protection	Check: • Damp proof course (position and potential damage). • Copings. • Projections (roof overhang, string courses). • Sills. See the chapter on Damp for more detail.
Vegetation	Identify any vegetation that may affect the durability of the wall. Some lichens, algae and mosses can cause minor surface damage and may also delay evaporation. Buddleia and ivy can cause significant damage to the mortar joints and even the masonry itself.

TIMBER FRAME CONSTRUCTION

INTRODUCTION

Modern timber frame construction is virtually indistinguishable from its brick and block counterpart. In the 1960s a number of houses, mostly local authority-owned, were built in one form of timber framing or another. Many of these were built in cross-wall construction, in other words with blockwork party walls and timber framed front and back walls. In recent years, timber framing has become quite popular partly because it offers several advantages over more traditional forms of building. These include:

- fast construction
- quick return on borrowed capital
- less dependence on traditional 'wet' skills
- reduced drying-out time due to elimination of wet trades
- reduced dead-load
- improved quality control
- less reliance on non-renewable resources.

In addition, timber frame construction can easily achieve high levels of thermal insulation, resulting in capital savings in heating equipment and lower costs-in-use.

The external walls are formed in preservative-treated timber and comprise a series of timber studs nailed to top and bottom timber plates. The outer face is covered with plywood sheathing or other suitable sheet material to stiffen the frame and help prevent deformation caused by wind-loading. A vapour permeable, waterproof layer protects the sheathing. The inside face of the panel is covered with plasterboard. The frame can be clad with timber weather boarding or masonry. Insulation is fitted between the studs. In some examples, it can also fill the cavity.

The panels are usually factory-fabricated and delivered to site, as and when required, ready for assembly.

Structural elements

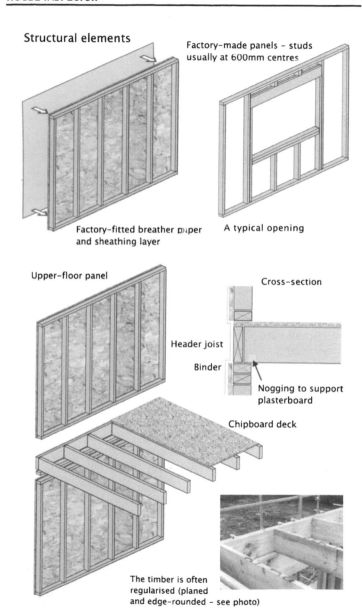

Factory-made panels - studs usually at 600mm centres

Factory-fitted breather paper and sheathing layer

A typical opening

Upper-floor panel

Cross-section

Header joist

Binder

Nogging to support plasterboard

Chipboard deck

The timber is often regularised (planed and edge-rounded - see photo)

DETAILS

Timber framing can be built on traditional foundations. The sole plate (see Timber frame – details on p52) is fixed to the ground floor slab or foundation blockwork by shot firing or with brackets. A DPC under the sole plate keeps it dry. The ground-floor frames are then nailed to the sole plate. If a timber ground floor is required the joists sit on the sole plate and form the base or platform for the wall panels.

First floor construction is shown on the previous page. A head binder nailed along the top of the panels helps tie them together. The joists sit on the head binder, nailed to a header joist running at right angles. The header joist prevents fire from entering the floor void. Floor boarding, usually in the form of chipboard or strand board, is laid right across the joists and forms the platform for the next lift of panels.

The roof construction is no different from any other house although where the frame is clad in brickwork allowance must be made for differential movement. The timber frame will actually shrink by about 15mm or so during the first few months (the shrinkage occurs across the grain and so all the timbers laid flat are affected, in other words the joists and plates). A small gap between the roof structure and the brickwork will prevent any damage.

The detailing around windows is somewhat tricky because of the need to prevent damp penetration, guard against fire spreading into the cavity, while at the same time allowing for differential movement.

Internal walls (partitions) are normally formed in studwork with plasterboard both sides. Quilts can be added between the studs to improve sound insulation. Partitions usually contain raking braces to help stiffen the whole frame.

Timber frame – details

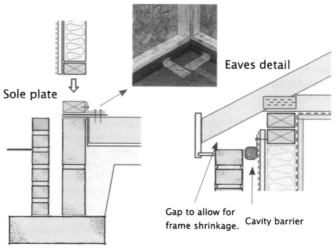

Sole plate

Eaves detail

Gap to allow for frame shrinkage. Cavity barrier

Head detail

Cavity tray over lintel to prevent damp penetration.

Weepholes for drainage.

Breather paper lapped over cavity tray.

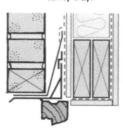

Lintel fixed back to frame with clips. A small gap allows for vertical frame shrinkage.

Jamb detail

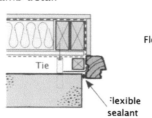

Tie

Flexible sealant

Typical sill detail with tiled subsill

Flexible sealant

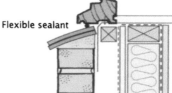

FINISHING THE PANELS

The insulation is normally mineral wool or glass fibre. The thickness obviously depends on the size of the studs – it can range from 80mm to 140mm. The inside face of the panel is usually lined with 12.7mm or 15mm plasterboard, with the joints taped and filled.

A vapour control layer (sometimes called a vapour check or vapour barrier) must be provided between the insulation and the plasterboard. This prevents condensation occurring within the wall panel. If moisture, in the form of vapour, passes through the plasterboard and through the insulation, it can condense on the inside face of the cold sheathing. The most common material is polythene; an alternative is to use a plasterboard that incorporates a vapour control layer – usually bonded to one face.

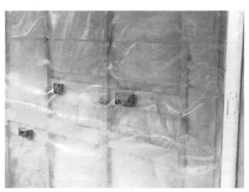

CLADDING

The external cladding in this country is normally masonry. Masonry, whether it be brickwork, stonework or blockwork, requires its own foundation and it must be tied back to the wall panels at regular intervals using special wall ties. The wall is quite thin and without restraint it would be unstable.

An alternative approach is to clad the panels in timber. This is the norm in North America and Scandinavia. It is cheap, quick and light. Also it avoids problems of differential movement. It does not, however, seem to be popular with the public.

Timber frame – cladding

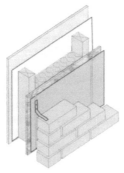

The ties must be fixed to the studs, not just the sheathing; hence the factory-fitted thin blue tape.

In the UK most timber framed buildings are clad in brickwork. A 50mm or 75mm cavity is normal. The brickwork is tied back to the timber framing with flexible ties (to allow for differential movement).

Cavity barriers (see Fire Protection) must be correctly positioned before the brickwork starts.

The breather paper protects the frame during and after the construction process.

Timber framing can also be clad in various types of timber boarding. These are not very popular in the UK but they are cheaper and quicker.

FIRE PROTECTION AND SOUND INSULATION

Separating walls

Where there are party walls, in other words in semi-detached and terraced houses, the separating walls must provide adequate fire protection and good sound insulation. The required standards can be achieved by using two sets of wall panels with a cavity (usually 50mm) between them. Adequate fire protection is provided by two layers of plasterboard fixed to the outer face of the frames (31mm minimum per side), and good acoustic performance is obtained by the thick plasterboard, the separation of the panels, plus an anti-reverberation quilt somewhere within the wall.

To improve the stability of the panels, light-gauge metal ties can be fixed at centres not more than 1.2m at, or about, first-floor level.

Fire stops are necessary to seal imperfections of fit at fire-resisting elements. They must be made from non-combustible material, such as mineral wool quilt, and are required at the junction of the separating wall and the external wall and at the junction of separating wall and roof.

Cavity barriers

To restrict the possible spread of fire or smoke in concealed places within the construction, cavities in timber structures must be closed with cavity barriers. A cavity barrier is a device that can be constructed of any material that is capable of providing fire resistance of at least half an hour and is typically made from a strip of mineral wool encased in a polyethyline tube. Cavity barriers must be fixed around the eaves and verge and are also required in the cavity to prevent fire from spreading to adjacent dwellings. Current Building Regulation requirements for cavity barriers are shown on the next page. This is in addition to the fire-stopping mentioned earlier, which keeps fire out of the separating wall (party wall). Until the 1990s, requirements for England were virtually the same as the current requirements for Scotland.

Timber frame – fire protection and window fixings

The party wall normally comprises two frames with a cavity in between. The boarding at one end is to provide racking resistance.

This is a block of four flats – the (red) cavity barriers slow down the spread of fire along the cavity. The quilt (inset) is party-wall fire stopping. This runs up the wall and up to the ridge.

The current requirements (2005) for the positions of cavity barriers in two-storey housing are shown as dotted lines – Scotland and Northern Ireland (left), England and Wales (right).

Windows are normally fixed before the brickwork commences. The batten (left) provides a fixing for the window and acts as a cavity barrier. The right-hand picture shows a window 'sleeve'. This provides a template for the bricklayers; the window itself will be fixed when the brickwork is complete.

DEFECTS

There are a variety of defects that have occurred in timber frame housing, most of which have been caused by poor supervision and bad site practice. However, a number of reports have concluded that timber frame housing is no less reliable than its traditional counterpart. Furthermore, most developers and architects are now aware of the potential pitfalls and modern timber frame housing should, if properly supervised, be durable. This last part of the chapter emphasises those areas of construction where particular care is required to prevent the risk of premature failure.

Omission or cutting of the vapour control layer

The vapour control layer should be at least 500 gauge polythene (or equivalent) and must be fixed to the warm side of the insulation. Fixing should be delayed until the frame has a moisture content of 20% or less. Any holes for services should be neatly cut and all tears repaired with special tape. Torn vapour control layers can increase the risk of condensation within the panel.

Unauthorised notching and cutting of the studs

During construction, the structural members are sometimes cut or notched to receive services. This can weaken the structure and may expose inadequately treated timber.

Base structure

It is important that the base structure is 'square' to avoid any problems of frame assembly and fixing. Although this is unlikely to be of structural significance, it might result in narrowing the cavity between masonry cladding and sheathing, with increased risk of rain penetration.

Wall panels

To maintain an even cavity with the brickwork, it is essential that the panels are erected vertically. Too narrow a cavity may result in damp penetration, too wide a cavity will lead to loose fitting cavity barriers. When the panels are in position, care is needed in placing/repairing the breather paper. It should be fixed with generous laps, and secured at regular intervals to prevent damage by the wind. The position of the studs should be indicated by chalk or tape to facilitate correct wall-tie fixing. A brick or block cladding must be fixed back to the frame with flexible wall ties and there should be adequate provision for differential movement. The timber frame will shrink by as much as 15mm during the first year or so, whereas new clay brickwork may even expand slightly.

Cavity barriers and fire–stops

There have been a few problems with incorrectly installed cavity barriers and fire-stops. In some cases they have been omitted completely and in others they have been of insufficient size to fill the cavity. As with the vapour barrier they are difficult to check once the building is complete, and therefore depend on well-trained operatives and good supervision.

Protection

Incomplete treatment of the timbers, usually as a result of poor quality control, can increase the risk of rot occurring in the finished structure, and inadequate weather protection on site can result in distortion as the frame dries and permanent weakening of chipboard floors. Incorrect temporary storage and poorly fixed breather paper is likely to result in damp timber.

Photographs

The photographs opposite show a number of potential defects.

1. The nails, which should be securing the sheathing to the studs, have missed their target.
2. The panels should be safely stacked and protected from rain.
3. The studs forming internal walls normally include raking braces to help stiffen the frame. They must not be omitted.
4. The chipboard or strand-board flooring normally runs right across the joists. This means that wall panels are supported on the chipboard. This makes renewing the boarding very difficult.
5. As point 4 above.
6. Before the brickwork commences, the breather paper hems should cover the bare timber at the quoin and lap the pre-cast ground beam.
7. Any torn breather paper should be repaired before the brickwork is built.
8. Wall ties should slope outwards slightly; and certainly not slope inwards. If this timber framing shrinks slightly this tie will fall (slope) the wrong way. Note the ties must be fixed to the studs, not just the sheathing.

Potential defects

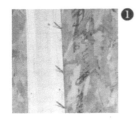

CHECKLIST

Element	Comments/problems
Is it timber framed?	Identifying timber frame construction (when brick clad) can be quite difficult. There are three or four ways to check without the need for any 'opening-up'. The windows are normally fixed to the timber frame, not the brickwork cladding. You are therefore likely to see the full depth of the brick reveal (looking at the window from the outside). In addition there should be a mastic-filled gap around the frame (to allow for differential movement). In the roof space the finish of the party wall (plasterboard) and the construction of the gable wall (timber panel) both indicate timber frame construction. From inside the property, it may be possible to lift the floor decking where the joists run into the wall to reveal the joist's support. In the 1960s and 1970s cross-wall construction was fairly common. This typically comprised party and gable walls in block or brick, with front and rear walls in timber framing.
External inspection (brick cladding)	The brickwork should be plumb and should be free of bulges or bowing. These can be caused if, at the construction stage, there was inadequate provision for shrinkage of the frame. Provision for movement is required around the windows and at the eaves. Uneven gaps around windows (at the jambs) may suggest the wall is bowing. Bowing can also be caused by inadequate restraint of the brickwork. There should also be weepholes above cavity trays to keep the moisture content of the frame as low as possible. Weepholes should be sited over window and door openings and in pre 1990s houses at first-floor level (above the horizontal cavity barrier). Weepholes are also normally required to drain the bottom of the cavity.

Element	Comments/problems
Internal inspection	Inspect around window openings for dampness. It could be penetrating dampness if vertical DPCs have been omitted or condensation (caused by cold bridging) if gaps between window and brickwork have not been properly filled. Look for pattern staining or mould growth on the internal face of the external walls. This suggests that areas of insulation have been omitted.
Roof, ground and upper floors	See appropriate chapters.
Opening up the construction	There may be situations where evidence of dampness requires further investigation. In this situation an optical probe can be used to inspect the various layers in the wall. Typical problems include: • mortar droppings in cavity • lack of cavity trays • retro-fitted cavity insulation • lack of weep holes • torn breather paper • torn vapour check • omission of vertical DPCs at jambs. It may be necessary to take meter readings of the softwood framing members (see chapter on Damp) if dampness is suspected or confirmed during the inspection. To do this some of the plasterboard or cladding will have to be removed. In a house that is over two years old readings should not exceed 22% (June to September) or 24% (October to May).
Fire protection	Establishing the adequacy of the fire protection is virtually impossible without opening up the construction. However, an optical probe will enable a partial examination of the position and condition of the cavity barriers and fire stopping. Note that the requirements for cavity barriers were revised in the mid 1990s (see main text). A probe will also enable an examination of the header joist, which is designed to prevent a fire in the cavity entering the upper-floor void.

GROUND FLOORS

INTRODUCTION

By the end of the 19th century, most new houses in towns and cities were built with timber ground floors. Imported softwood (deal) accounted for most of the timber; only a small percentage was home-grown. The floors were constructed from a series of timber joists supporting square-edged floorboards. The design and construction of timber floors has changed over the past 100 years but only in detail, not in principle. Modern timber floors are still constructed from softwood joists but nowadays there are stricter rules (in the form of Building Regulations) to prevent risks of damp and rot. In fact, timber floors are fairly rare in modern construction because most ground floors are made from concrete.

Concrete floors were not uncommon in the late Victorian period although they were often confined to certain rooms. Large suburban houses, for example, might have kitchen and hall floors formed in concrete, possibly with a tiled finish. Not until the 1950s was it common to find the entire floor cast in concrete. These floors were often finished with wood blocks or a sand/cement screed covered with asbestos tiles – the fitted carpets we take for granted today were rare until the 1960s. A concrete floor is basically an insitu slab of concrete poured over a levelled and compacted base. The floor is supported by the ground beneath it, not by the house walls. Nowadays a damp proof membrane (DPM) should be included in the construction to prevent capillary action (rising damp).

Some conditions are unsuitable for insitu ground floors. These include wet sites (ie high water tables), ground containing aggressive chemicals, sloping sites, and sites with poor ground. In these situations, it is often more practical and economic to use suspended concrete floors. These do not 'sit' on the ground but are supported by the house walls. Today, most suspended floors are made from pre-cast inverted 'T' beams with a block infill.

TIMBER GROUND FLOORS

Most houses built before the 1950s have ground floors constructed from timber. They are usually referred to as suspended or raised timber floors.

The floor comprises a series of joists supported by external and internal loadbearing walls and covered with floorboards. The size of a joist depends largely on its span; as its length or span increases, so must its depth to safely support the load imposed upon it. Deep joists are expensive and to reduce joist size there are usually intermediate supports known as sleeper walls. These are small walls in rough stone or brickwork built directly on the ground or on small foundations. In practice, ground-floor joists are often half the depth of those used in upper floors where, of course, such intermediate support is not possible.

The drawings on the next two pages show the ground-floor layout of a typical late Victorian terraced house. The joists run from party wall to party wall (they often run front to back) and are supported mid span by sleeper walls. The joists, typically 100mm x 50mm, are usually at 400mm centres (16 inches) as this generally offers the most economic arrangement of the timbers.

To ventilate the sub-floor void terracotta or cast-iron air-bricks were built-in to the external wall. Towards the end of the Victorian period DPCs, often formed in brittle materials such as slate, were becoming common.

Although these floors can give sterling service there are some common problems:

- Ground level often changes so vents become blocked.
- Inadequate number of vents or poor vent positioning.
- Lack or failure of DPCs.
- Underfloor void floods (if below external ground level).
- Joist ends become rotten due to end penetration.

These are covered in more detail in a later section.

Late Victorian ground floor

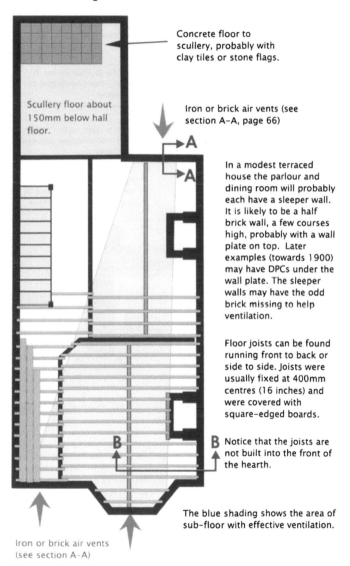

Concrete floor to scullery, probably with clay tiles or stone flags.

Scullery floor about 150mm below hall floor.

Iron or brick air vents (see section A-A, page 66)

In a modest terraced house the parlour and dining room will probably each have a sleeper wall. It is likely to be a half brick wall, a few courses high, probably with a wall plate on top. Later examples (towards 1900) may have DPCs under the wall plate. The sleeper walls may have the odd brick missing to help ventilation.

Floor joists can be found running front to back or side to side. Joists were usually fixed at 400mm centres (16 inches) and were covered with square-edged boards.

Notice that the joists are not built into the front of the hearth.

The blue shading shows the area of sub-floor with effective ventilation.

Iron or brick air vents (see section A-A)

65

Late Victorian ground floor

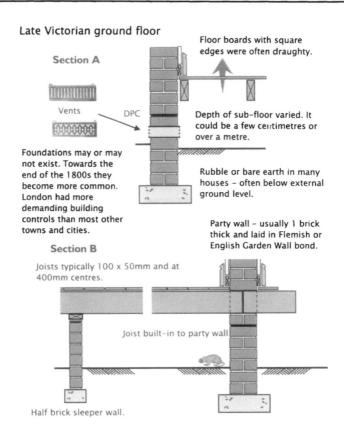

Section A

Vents

DPC

Foundations may or may not exist. Towards the end of the 1800s they become more common. London had more demanding building controls than most other towns and cities.

Floor boards with square edges were often draughty.

Depth of sub-floor varied. It could be a few centimetres or over a metre.

Rubble or bare earth in many houses – often below external ground level.

Party wall – usually 1 brick thick and laid in Flemish or English Garden Wall bond.

Section B

Joists typically 100 x 50mm and at 400mm centres.

Joist built-in to party wall

Half brick sleeper wall.

In the late 1920s these floors were improved in an attempt to deal with the problems highlighted above.
Note that:

- The entire floor is separated from the substructure by the DPCs.
- The bare earth is covered with a concrete slab (often referred to as 'oversite'), which is at, or above, external ground level to prevent the build up of water. The slab also prevents growth of vegetation.
- The floor joists are supported by honeycombed sleeper walls, through which air can pass easily, and the joists do not touch the external wall.
- Vents are provided, well clear of ground level, and sleeved to

prevent cold air entering the cavity. The requirements of the current Building Regulations are quite onerous regarding the amount of ventilation and this obviously reflects concern about this form of floor construction.

- As most of these houses were semi-detached or detached, the underfloor void is relatively easy to ventilate.

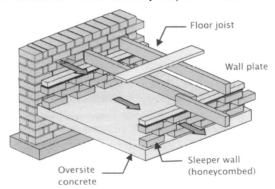

Floor joist

Wall plate

Oversite concrete

Sleeper wall (honeycombed)

Until the 1970s, floorboards were always made from timber, and nearly always softwood. Square-edged boards were gradually displaced by tongued and grooved boards in the years either side of the Second World War.

Square-edged boards   Grooved or rebated boards

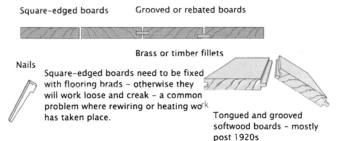

Brass or timber fillets

Nails

Square-edged boards need to be fixed with flooring brads – otherwise they will work loose and creak – a common problem where rewiring or heating work has taken place.

Tongued and grooved softwood boards – mostly post 1920s

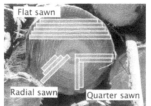

Flat sawn

Radial sawn   Quarter sawn

Flat sawn boards are more likely to twist or curl and can 'shell-out' if laid upside down.

Modern raised timber floors can be built in a number of ways. The oversite can be a 100mm concrete slab or a 50mm layer of concrete blinding on a DPM. The joists can be supported on sleeper walls or on joist hangers. Insulation (now required) can be laid between the joists, or above them if it is a rigid material.

Part of a modern raised timber floor. Ventilation requirements are now quite onerous – about 1 air brick, 225 x 150mm, every 1.5 metres.

Most modern timber ground floors are covered in chipboard or strand board. If the boards are not fixed properly there are a number of potential problems. Most of these are summarised in the next secton.

Insulation

Although a timber floor is perceived as being warm, it does permit considerable heat loss due to the flow of air underneath the joists. Although insulation would never have been provided during the original construction, today's ever-increasing fuel prices and emphasis on energy conservation makes it a sensible option. Since 1990 it has been a requirement that all new ground floors contain insulation. Two simple methods are shown in the following diagrams and are suitable for both new and existing floors.

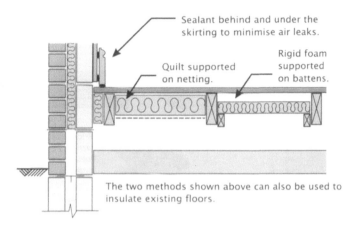

Sealant behind and under the skirting to minimise air leaks.

Quilt supported on netting.

Rigid foam supported on battens.

The two methods shown above can also be used to insulate existing floors.

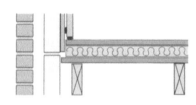

Insulating an existing floor without removing any of the boards is another option although it will normally entail removing the skirtings. Most insulation boards cannot be laid directly on top of the joists – hence the construction shown here.

CONCRETE FLOORS

Concrete ground floors were not unknown in the 1930s: they became more common in the 1950s because of the post-war restrictions on imported timber. The floor is basically a bed of concrete, supported by the ground directly beneath it, and quite independent of the surrounding walls. This type of floor was the most common from the 1950s until the 1990s.

Typical construction

A typical floor from the 1950s (there are earlier examples) might comprise a layer of hardcore (stone or broken brick), a concrete slab probably 100–125mm thick and the floor finish. This is often timber to disguise the nature of the floor, or in cheaper construction thermoplastic tiles laid in bitumen adhesive. Some floors, by no means all, contained damp proof membranes, usually liquid based.

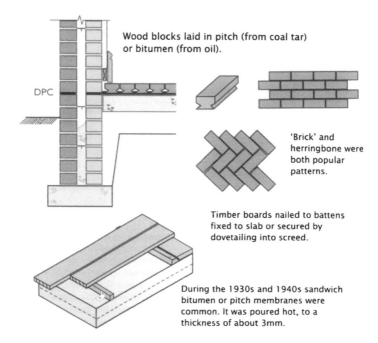

Wood blocks laid in pitch (from coal tar) or bitumen (from oil).

DPC

'Brick' and herringbone were both popular patterns.

Timber boards nailed to battens fixed to slab or secured by dovetailing into screed.

During the 1930s and 1940s sandwich bitumen or pitch membranes were common. It was poured hot, to a thickness of about 3mm.

In many houses with wood blocks or thermoplastic tiles the only barrier to rising damp was the bitumen bedding material.

Thermoplastic tiles were first produced in the UK just after the Second World War. The tiles were made from a mixture of resin binders, mineral fillers, asbestos and pigments. Most were 9 inches square (225mm). Early tiles were quite brittle. Asbestos vinyl tiles were introduced in the mid 1950s; they were made in much the same way but they were more flexible.

Typical construction of the late 1960s to 1980s

In the late 1960s a typical floor would comprise a layer of hardcore, a polythene damp proof membrane laid on a thin bed of sand (to prevent puncturing), and a floor screed probably covered with vinyl tiles.

Hardcore

The hardcore prevents loose earth from contaminating the concrete, helps spread the load evenly, makes up levels and reduces capillary action. The hardcore should be chemically inert and should be compacted in layers (see section on Floor Failure).

Damp proof membrane

In the mid 1960s polythene damp proof membranes (DPMs) became an accepted form of damp proofing. This barrier was usually laid below the concrete slab. DPMs on top of the slab, ie sandwiched under the screed, were also common and usually in liquid form, eg hot bitumen, or cold bitumen in solution. Liquid DPMs gave the best protection but were more expensive.

Concrete slab

The concrete slab was usually 100–125mm thick. In certain situations, ie where the ground was uneven or where there were soft spots below the slab, it might be reinforced with a mesh.

Floor screed

The function of the floor screed is to provide a smooth finish suitable for carpets or tiling. The screed is laid towards the completion of the building prior to hanging the doors and fixing the skirtings. It is a mixture of cement and course sand (typically one part cement to three or four parts sand) mixed with the minimum amount of water and laid to a thickness of 38–50mm. Excess water will weaken the screed and delay the subsequent laying of any floor tiles.

Typical construction from the 1970s and 1980s

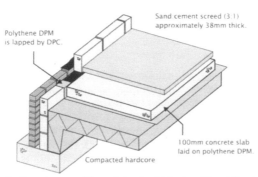

Sand cement screed (3:1) approximately 38mm thick.

Polythene DPM is lapped by DPC.

100mm concrete slab laid on polythene DPM.

Compacted hardcore

Liquid membranes could be applied on the slab (in place of the polythene).

Insulation

The Building Regulations now require insulation in ground floors. A variety of manufacturers produce a range of rigid insulation boards which can be laid above or below the slab. Some of the boards have a closed cell structure and are impervious to both water and vapour. They can therefore be laid under the DPM (the DPM is still necessary to prevent moisture rising between the board joints and penetrating the slab). Where boards are laid under the DPM the blinding referred to earlier is not always necessary. Typical construction is shown below.

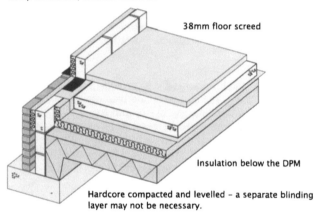

Note: not all insualtion boards are suitable for use in damp conditions, ie below the DPM.

38mm floor screed

Insulation below the DPM

Hardcore compacted and levelled – a separate blinding layer may not be necessary.

Chipboard flooring

In modern construction, chipboard floating floors have become a common alternative to sand/cement screed. Chipboard and strand board are both very sensitive to moisture and a vapour control layer is normally required under the boarding to prevent drying construction water (in the concrete) affecting the floor. This membrane is in addition to the DPM below the slab. The boarding is laid with glued tongued and grooved (t&g) joints and with a perimeter gap of 10mm or so to allow for moisture and thermal expansion. This gap is covered by the skirting.

Moisture resistant chipboard or strandboard laid on vapour control layer. The VCL protects the chipboard against moisture in the slab (construction water drying out).

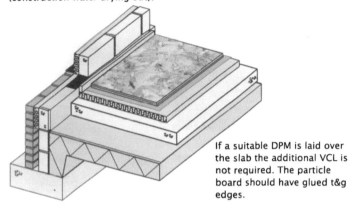

If a suitable DPM is laid over the slab the additional VCL is not required. The particle board should have glued t&g edges.

Suspended concrete floors

In certain conditions the use of a normal concrete slab is unsuitable:

- In shrinkable clays.
- Where a large volume of vegetable soil or soft unsuitable ground has to be excavated.
- On steep sloping sites where the hardcore depth will be uneven.
- Where the water table is high.
- Where there are aggressive chemicals in the soil.

In these situations it is preferable to use a suspended concrete floor. This is supported by the external and internal loadbearing walls and is independent of the ground beneath it. Should any settlement or heave of the ground beneath it occur, it will not be affected. The floors are usually made from a series of inverted 'T' beams, 150–200mm thick, with a concrete block infill. When the walls are built to the appropriate level and the DPC is positioned, the beams can be manhandled or craned into place. Concrete blocks are then positioned in between the beams.

When the beams and blocks are in position a moist mix of cement and sand is brushed over the surface to fill any gaps. The floor is then ready to receive the floor finish, which will be laid towards the completion of the building. The two most popular finishes are screed or particle board, both laid on insulation.

A beam and block floor showing the periscope-type vents.

Nowadays the Building Regulations require that the underfloor space is vented, before 2004 it was only necessary to ventilate the space if the ground was not well drained or if there was a risk of gas build-up.

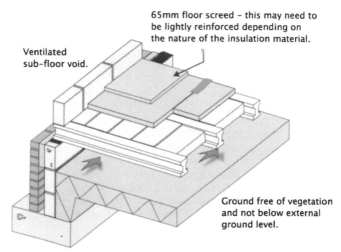

65mm floor screed - this may need to be lightly reinforced depending on the nature of the insulation material.

Ventilated sub-floor void.

Ground free of vegetation and not below external ground level.

The insulation should have taped joints; some manufacturers prefer a slip layer laid right across the insulation (building paper or thin polythene).

FLOOR FAILURE

Timber floors

Until the early 1900s, it was common to build joists into the external wall to provide end support. This increases the risk of moisture penetrating the end grain of the joists. Some properties have slightly thicker walls below floor level and this gives improved protection. Until the beginning of the 20th century, damp proof courses were rare in the majority of houses. This means that rising damp can not only attack the joist ends, but also the middle of the joists where they sit on the intermediate sleeper walls.

The level of the earth or rubble infill under the floor is often below external ground level and in wet conditions, or areas with high water tables, the underfloor space can often be permanently damp.

To keep the underfloor spaces relatively dry they were ventilated. This was usually achieved by a series of cast-iron or terracotta vents positioned just below floor level. Over the years the vents can become blocked, either through changes in external ground level or just general accumulation of debris. The reduced ventilation can lead to excessive levels of damp, which, ultimately, can cause outbreaks of rot. In addition, there are usually not enough vents to do the job properly. Have a look at the first drawing in this chapter. It should be clear which parts of the floor are most likely to suffer from rot.

Sleeper walls also reduced ventilation through the floor if they were not fitted with vents or honeycombed.

Due to the limitations of woodworking machinery square-edged boards were common until the 1950s. Once in position the boards would shrink slightly as their moisture content stabilised, resulting in small gaps appearing between the boards and uncomfortable draughts in the downstairs rooms. Twisting boards can also damage floor finishes. This is a particular problem where boards have been removed and re-nailed incorrectly. Cut floor brads are the best form of nailing.

The two left-hand photos on page 76 show the ground floor of a terraced house built in 1903. A recent inspection highlighted a number of potential problems with this floor.
- There were not enough vents (only one at the front and three at the rear).

- The joists were built in to the external walls – although they were sitting on a DPC.
- The surface of the ground had been blinded with concrete although its surface level was 100mm or so below external ground level.
- The sleeper walls had some perpend joints missing to encourage ventilation but did not have a DPC below the wall plates.

The floor in the photo (right) was built in 1959. According to the owner (who has lived in the house since it was built), the water has always been a feature. If you look at the bottom three courses of bricks you can see a tide mark. In wet conditions the water rises. The floor itself has not been affected. The absence of a wall plate on the sleeper wall is unusual.

Rot

Wet rot is most likely to occur where the joists are built into the external walls. Dry rot can also occur in timber floors. (See chapter on Rot and Insect attack.)

Sizing floor timbers

To estimate the adequacy of joist sizes in older buildings there is a simple rule of thumb – divide the span in mm by 25, and add 50mm. This assumes the joists are at 400mm centres and that they are 50mm wide (they usually are). 400mm is the metric equivalent of 16 inches (the traditional centres). For a 2000mm span the joists should be at least 130mm deep. In imperial measurements joists were usually available in 25mm increments (1 inch), ie 150, 175, 200, 225, 250mm. Anything over 250mm (10 inches) is rare. Most ground floors will have supporting sleeper walls and the joists are unlikely to exceed 150mm in depth. Remember the span is between sleeper walls.

These rule of thumb methods are a rough guide only. A modern floor will be subject to the Building Regulations. Tables in the Regulations (since 2004 these have been published by The Timber Research and Development Association – TRADA) give joist sizes for specific types of timber and specific loads. However, remember the Building Regulations are not retrospective.

Modern timber floors

In modern raised timber floors none of the defects described in the previous page should occur. However, there are other problems; in the main these are caused by poor workmanship when laying the boarding. The main problems are summarised on the right.

Creaking boards (new and old floors) can also be caused by re-fixing boards over quilt insulation.

Modern floor fixings

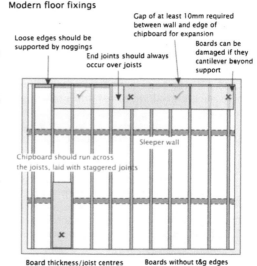

Loose edges should be supported by noggings

End joints should always occur over joists

Gap of at least 10mm required between wall and edge of chipboard for expansion

Boards can be damaged if they cantilever beyond support

Sleeper wall

Chipboard should run across the joists, laid with staggered joints

Board thickness/joist centres
18mm up to 450mm centres
22mm 600 centres
Moisture-resistant chipboard is more durable than standard floor grade.

Boards without t&g edges will require noggings under both long sides.
Wire nails can work loose and will damage floor coverings; ring shank nails are the correct ones to use.

SOLID FLOORS

Damp penetration

Where a floor slab is protected from rising damp by a polythene DPM laid beneath the slab, a variety of problems can occur, most of them caused by poor site practice. Polythene is available in rolls 5m wide – where the rooms are wider than this the polythene will require jointing. Joints can be formed by either using special tape or by folding the polythene to form a welted seam.

This is fine in theory but, in practice, on a wet and windy building site in the middle of winter this is not so easy to achieve. More often than not, the polythene is just lapped (without the tape), and this can permit damp penetration. Other problems can occur as the concrete is poured. Wheelbarrows and operatives' boots can puncture the polythene even if it is laid on a sand blinding.

If the hardcore is not well compacted, or if soft ground beneath the slab has not been removed, the slab can settle slightly. If this movement is excessive the DPM can tear at the edges and creates a path for damp to penetrate the superstructure.

Where a liquid DPM is used above the slab, problems can occur if the concrete is not thoroughly cleaned before it is applied. All traces of dust, plaster and mortar should be removed prior to applying the liquid if it is to bond successfully. It is also important to ensure that there is a good link with the horizontal DPC in the internal leaf of the cavity wall.

Many houses, especially those built in the 1950s and early 1960s were not built with DPMs. They often had wood blocks or thermoplastic tiles as a floor finish. The bitumen adhesive partly acted as a DPM. These may still perform quite adequately, particularly where the house is well ventilated. However, problems can arise where the blocks or tiles are removed or where the house is extensively draft proofed. Where vinyl sheet is placed over floors without effective DPMs its underside soon becomes wet.

Damp patches around the edge of a floor may be caused by a failure of the DPM/DPC joint but also can be caused by condensation. This can be through condensation run-off from windows and doors, but also from condensation occurring around the edge of the slab.

Ground movement

Settlement caused by poor compaction of the hardcore is common; where it is minor it can be ignored. Over the years settlement can be substantial – usually manifesting itself first with gaps below skirtings. There are other causes of settlement that can have more significant effects. Some clay soils expand and contract due to changes in moisture content. When building on ground of this type it is important that the foundations are at a depth where the ground is stable (piled foundations are often necessary), and that a suspended concrete floor is used. Some silts and sandy soils can also expand and this is usually caused by frost attack. As the water in the soil turns to ice the ground expands and lifts the slab. It usually only occurs at shallow depths and should not affect foundations. It is also comparatively rare in housing as the ground beneath the slab is prevented from freezing due to heat escaping through the floor. It is, however, more common in unheated buildings, such as garages, and may become more commonplace due to the recently introduced thermal requirements for floors, which obviously reduce the amount of heat escaping into the ground.

Chemical attack

The most common form of this is sulfate attack caused by sulfates in the hardcore (and sometimes the ground below) reacting with the concrete and forming a compound which expands. In damp conditions the sulfates in solution can permeate the concrete, thus allowing a chemical reaction to occur. As the chemical reaction occurs the slab lifts and cracks (it can even push the walls outwards slightly). If the problem has occurred, repair is very expensive and will probably entail complete renewal of both the hardcore and the slab.

Some hardcores are affected by water. As they become wet they expand and lift the slab.

Both these problems can lead to arching and doming of the slab – this can easily be checked with a straight edge. If the problem is sulfate attack the floor will lift clear of the hardcore. If the hardcore has expanded both the floor and hardcore will remain in contact. Obviously, drilling through the floor is the only way to check this.

Concrete failure

Sulfate attack is probably the most dramatic way in which concrete can fail but, in practice, poor site supervision can also result in a variety of problems, which can be expensive to resolve. As mentioned previously, concreting in adverse weather can cause problems. There is also the danger of using cement that is excessively old, incorrect mixing of materials and contaminated aggregates. It is common practice on site to add extra water to the concrete before it is poured. Although this makes it easier to place, the excess water will eventually evaporate and create tiny voids in the concrete. This seriously reduces its strength.

Screed failure

The two major causes of screed failure are inadequate thickness and material failure. Screeds should normally be 38–50mm thick if there is a good bond with the slab, but up to 75mm thick if a polythene membrane and/or insulation is used under the screed. Material failure includes incorrect mix of materials, contaminated sand, old cement or excess water. In both cases, the only solution is to break up the screed and relay it. This is an expensive operation, particularly if skirtings need refixing and doors need trimming and rehanging.

To prevent premature failure of bonded floor coverings such as vinyl tiles, it is vital that the screed is dry before the covering is fixed. Failure to do this is likely to result in lifting of the tiles.

Floor finishes

Failure of wood block floors can be caused by:

- Wetting (leaking washing machines are the worst culprits).
- Shrinkage (joints open up in the winter when homes are heated and close up during periods of warm weather with high humidity).
- Minor slab movement causing cracking of the brittle bitumen bedding.
- Incorrect polishing (ie, wrong material) can fill the joints preventing thermal movement. This can lead to arching of the floor.
- Long-term breakdown of bitumen bedding (over the years it becomes brittle).
- The blocks can wear badly because they are often cheap softwood.

Thermoplastic tiles can give sterling service. However, there are a few common defects.

- Edges can become damaged as furniture is dragged across a room. The tiles are quite brittle and edges are easily snapped off.
- In wet conditions soluble salts rising through the slab can evaporate, leaving behind bands of a white crystalline material around the edges of the tile. Over time, the tiles delaminate.
- Long-term breakdown of bitumen bedding (over the years it becomes brittle).

Ceramic tiles are most affected by loss of grouting and loss of bond. If the grout is too weak (contains too much water) or if it does not fill the full depth of the joint it can fail prematurely. Loss of bond (sometimes manifesting itself as fairly dramatic arching) can be caused by substructure or mortar shrinkage. It can be exacerbated by slight moisture expansion in the tiles after they are laid.

Chipboard and strand board can disintegrate if they become wet. Even moisture resistant grades can de damaged where they are in long-term contact with moisture. The boards can swell and crown if they have been laid without adequate perimeter gap.

Suspended concrete floors

These floors are unlikely to fail as they are not in contact with the ground and are produced in conditions where it should be possible to guarantee quality control. However, they are a comparatively new form of construction and unforeseen problems may occur in the future. It is interesting that the 2004 Building Regulations, Part C, requires these floors to be ventilated in all situations. The life of the pre-stressed beams is, as yet, unknown.

CHECKLISTS

Material	Identifying feature
Establish whether the floor is suspended timber, suspended concrete, or a ground bearing slab.	The starting point is probably the age of the property (although the floors may have been changed). Until the 1950s most ground floors were suspended timber. Check for external air-bricks and, more importantly, the feel of the floor itself. From the 1950s to the 1990s most houses were built with ground-bearing slabs. Some of these will be finished with floor screeds, others with floorboards on battens or chipboard on insulation. Most ground floors, nowadays, are made from suspended pre-cast concrete beams with a block infill. These may be ventilated and can be finished in all the ways described above.

TIMBER FLOORS

Element	Comments/problems
Adequate ventilation and damp proofing (viewed from outside).	Are there enough vents, are they adequately positioned (to allow cross-ventilation) and unblocked? Is there a DPC visible, well clear of ground level? (The DPC may be above or below the vents.)
From inside the room...	If the room feels or smells damp, rot should be suspected and further investigation will be required.
Structural stability	The floor should feel firm when walked over. Rattling windows or cabinets requires further investigation. Gaps, exceeding 10mm or so, between the skirtings and the boards can lend further support to this course of action.
If the floorboards can be lifted...	Check the following: • Integrity of the joists, particularly near the ends. • Joist size and span. • Damage caused by splitting, alteration work, building services installations.

Element	Comments/problems
If the floorboards can be lifted...	• Rot and insect attack. • Sleeper walls — including provision for ventilation, DPCs and wall plates. • Depth of void and risk of flooding; clear of debris. • Nature of the ground surface (eg bare earth, blinding, or concrete slab). • Is the floor insulated (and has the insulation been correctly installed)?
Floorboards	Check the following: • Adequate thickness. • Excessive gaps between floor boards. • Integrity of the boards – ie, free from splitting, twisting and cupping. • Creaking – usually caused by incorrect nailing or by fixing boards above insulation. • End joints (head joints) should be supported by joists.

GROUND–BEARING SLABS

From inside the house...	Try to identify the construction. Common combinations include: • Slab, screed and tiling. • Slab, screed and block flooring. • Slab, screed and carpet or sheet vinyl. • Slab, insulation and chipboard. • Power-floated slab. In recent years laminated interlocking flooring has become very popular. It can be laid over a screed or chipboard on a thin resilient layer. It is often a DIY job and may be covering all sorts of problems.
Initial floor inspection	The floor should feel level when walked over. Check the gaps below doors and skirtings. Any signs of dampness should be investigated.
Settlement of slab	Settlement of the slab is the most common problem. It is usually caused by poor compaction of the hardcore. Differential settlement is likely to result in cracking. In either case settlement is likely to damage the DPM where it joins the DPC.

Element	Comments/problems
Floor heave	If the floor seems to rise or crown the cause could be: • Heave of clay soils. • Expanding hardcore. • Sulfate attack. This may be accompanied by lateral expansion of the slab, which can push the external walls outwards. The DPC normally provides a slip layer. Any form of heave is likely to damage the DPM.
Floor screed	In most cases floor screeds will be covered with a floor finish and minor defects will not necessarily manifest themselves. If the screed is too thin or an inappropriate mix, it can crack. This can sometimes be felt as you walk over the floor. Screeds can also crack if they contain embedded pipes. Some screeds include service ducts. If these are not flush with the screed they can damage floor coverings.
Floor finishes	If the floor is covered in wood blocks or thermoplastic tiles see the main text for typical defects. Chipboard floors (laid over insulation) can crown if thermal and moisture movement is restricted. If floor finishes are lifting also suspect dampness.
Damp patches	Damp patches on a floor can be difficult to diagnose. Consider: • Leaks from washing machines or services. • Condensation run-off from windows. • Leaks around door thresholds. • Cold bridging around the slab perimeter. • Failure of the DPM or DPC.

SUSPENDED PRE–CAST FLOORS

Grout	These floors are relatively new and no significant problems have come to light. However, if the floor creaks when walked over it is likely to be caused by failure to brush grout in between the beams and blocks.

UPPER FLOORS

INTRODUCTION

Timber floors are usually constructed from a series of timber joists covered with timber floorboards or, nowadays, sheets of chipboard. Some developers favour the use of modern metal web joists and 'I' joists rather than traditional 'cut' timber. In modern construction, the size and spacing of the joists are subject to the Building Regulations. Before 1965 they were mostly controlled by Model By-Laws.

The depth of a joist affects its permitted span. The width affects its resistance to twisting and warping. Joists less than 38mm wide readily warp, and also provide insufficient area on which to nail and join floorboards. Joists less than 38mm wide should always be noted during a survey. Old joists are unlikely to meet the requirements of current regulations but may still be satisfactory. A rule of thumb method for assessing acceptable joist size can be found in the previous chapter.

EARLY FLOORS

The graphics on the next two pages show typical late Victorian/Edwardian construction for the floor of a modest terraced house. In summary:

- The joists are normally at 400mm centres (16 inches).
- Joists are often built-in to external walls although in better-quality construction they can be supported on iron brackets.
- Joists are also supported on internal loadbearing walls.
- Joists should not be built-in to party walls.
- The joists will normally run the shortest distance across a room.
- The joists are not built-in to hearths but trimmed around them.
- Strutting stops the joists from twisting.
- Most Victorian floors were originally covered with square-edged boards.
- Ceilings were usually timber lath and plaster (see Finishes on page 91).

Typical upper floor c1900

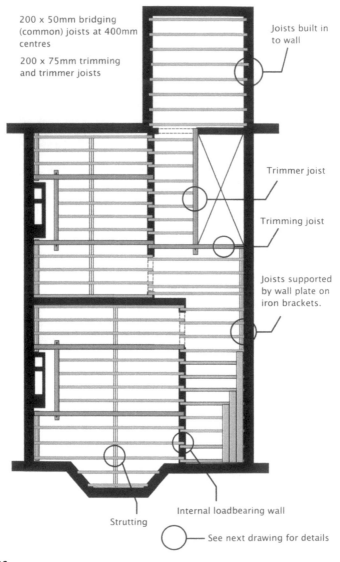

200 x 50mm bridging (common) joists at 400mm centres

200 x 75mm trimming and trimmer joists

Joists built in to wall

Trimmer joist

Trimming joist

Joists supported by wall plate on iron brackets.

Strutting

Internal loadbearing wall

See next drawing for details

Typical upper floor c1900 – details

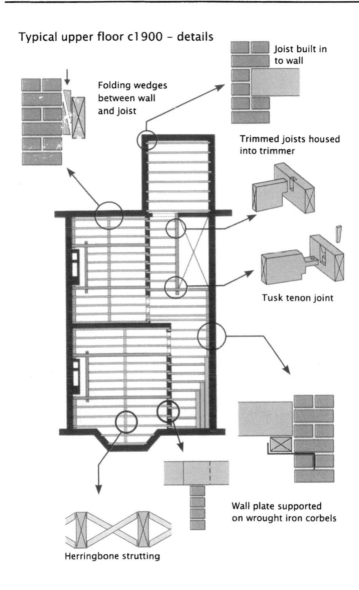

Folding wedges between wall and joist

Joist built in to wall

Trimmed joists housed into trimmer

Tusk tenon joint

Wall plate supported on wrought iron corbels

Herringbone strutting

Some larger Victorian and Georgian properties often have walls much thicker than 1 brick. In such properties, it was common to support the floors as shown in the diagram below. In this situation the joist and wall plate it sits on are protected by at least 1 brick thickness of wall.

Where rooms were large, double floors could sometimes be found. These often have one large-section beam supporting the joists in mid-span. In some of the grandest houses, sometimes there were a separate series of joists supporting the ceiling. This ensured minimum deflection and therefore minimum damage to expensive plasterwork.

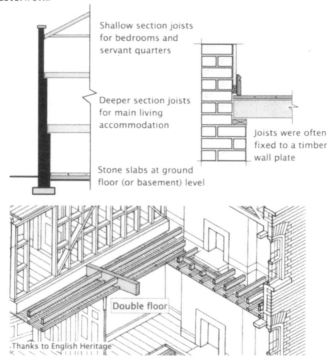

Shallow section joists for bedrooms and servant quarters

Deeper section joists for main living accommodation

Joists were often fixed to a timber wall plate

Stone slabs at ground floor (or basement) level

Double floor

Thanks to English Heritage

By the 1950s floor construction had evolved slightly:

- Walls in cavity construction offered better protection to the joists.
- Timber lath and plaster ceilings gave way to plasterboard.
- Tongued and grooved boards replaced square-edged boards.

MODERN FLOORS

The construction of modern upper floors is shown on the next two pages. They differ from 1950s floors in three main ways: strapping is now required to restrain the external walls, joist hangers have become common, and floorboards have largely been replaced by particle boards.

Typical modern floor

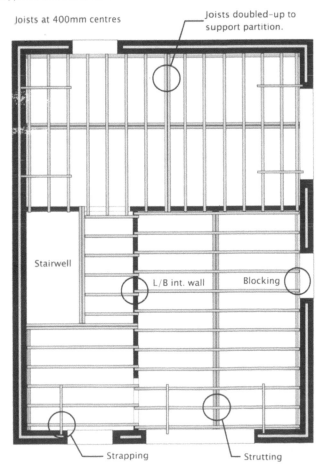

Joists at 400mm centres

Joists doubled-up to support partition.

Stairwell

L/B int. wall

Blocking

Strapping

Strutting

Typical modern floor – details

A modern upper floor is not very different from its Victorian counterpart. A floor still needs to be trimmed around an opening although nowadays metal hangers have replaced carpentry joints. Strutting is still required at right angles to the joists.

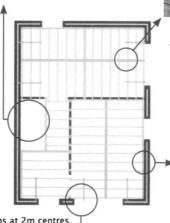

Joists can be built-in or supported on hangers. Blocking or noggings between joists are usually required for flooring and plasterboard edge support.

Straps at 2m centres.

Joists supported on standard hangers will require additional strapping (at 2m centres).

Restraint strapping – in this case above a window. Notice the blocking in between the joists (right).

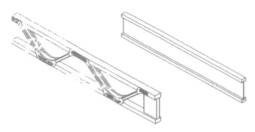

In recent years the use of metal web joists and 'I' joists has become more common. In principle these are no different from traditional 'cut' joists. One advantage is that they are capable of increased spans. The use of the metal web joists also precludes the need for potentially damaging joist notching.

Sound insulation

Since 2004 the Building Regulations has required sound deadening quilt (to reduce the passage of airborne sound) in upper floors (depending on room layout above and below the floor). In practice, of course, sound insulation of floors is improved by the use of carpets and good quality underfelt.

Finishes

Most modern floors are covered with tongued and grooved (t&g) chipboard or other particle board. It is still acceptable to use t&g floorboards, particularly where the floor is to be left exposed and varnished as a decorative feature. Modern ceilings are nearly always formed in plasterboard. This is covered in more detail in the chapter, on Finishes but the construction basically consists of large sheets of plasterboard, nailed or screwed to the underside or soffit of the floor, and finished with either plaster or some kind of flexible finish, such as Artex.

Fire protection

The purpose of fire protection is not to render a floor non-combustible, but to ensure that it will resist collapse for a sufficient period of time to allow evacuation of the occupants. In two-storey housing, the use of a layer of 12.5mm plasterboard (15mm is better) fixed to the underside of the joists provides adequate fire protection. Higher buildings have more onerous requirements with regard to fire protection and suitable forms of floor construction, which may include two or even three layers of plasterboard fixed to the joists, are contained in the Building Regulations.

Flats

In flats, the floors separate dwellings and, therefore, must provide good fire protection and resistance to the passage of impact and airborne sound. A detailed explanation of sound and fire protection is beyond the scope of this book. However, it is worth summarising typical construction. In houses that have been converted into flats the separating floors are likely to be timber. The quality of the sound insulation and fire protection will vary considerably depending on the date of the conversion. In good construction, there will be:

- Some form of sound deadening material in between the joists.`
- A barrier to smoke – usually in the form of t&g boards or a ply/hardboard overlay (on square-edged boards).
- A floating floor usually comprising sheets of chipboard bonded to a rubber or foam base.
- Additional layers of plasterboard to increase the mass of the floor and to improve fire protection.

Note that increasing the mass of the floor improves its resistance to airborne sound; adding a floating floor finish improves resistance to impact sound.

All this, of course, adds to the weight of the floor. In some houses floors have been converted into double floors (usually by adding steel beams) to cope with the increased load.

In new flats separating floors are most likely to be formed in pre-cast concrete. The floors are usually finished with a cement/sand floor screed laid on a resilient quilt or layer of foam. The soffit is normally finished with one or two layers of plasterboard fixed to timber battens. If two layers of battens are secured to the soffit, a service zone is created.

FLOOR FAILURE

End support

In houses with solid walls the joist ends are usually protected by approximately 100mm of brickwork. Minor defects in the wall (usually the pointing or a render finish) can result in damp penetration and potential problems of rot. Damp penetration can also occur in cavity walls and is usually caused by mortar droppings in the cavity or incorrect fixing of insulation. These problems can be exacerbated if the joists project into the cavity. See the chapter on Damp for more information.

In many modern houses joists are supported on hangers. The use of joist hangers helps the keep the joist ends dry. However, there are a number of potential problems associated with hangers. These include:

- Using the incorrect size or grade of hangers.
- Poor fixings between hanger and joist.
- Joists cut too short.
- Joists laid before hangers are loaded (two or three courses of blocks should be laid above the hangers before the joists are fixed).

Wall restraint

In modern construction wall restraint is a requirement of the Building Regulations. Joists supported by a wall can be built-in, supported on hangers with restraint straps every 2m, or supported by special restraint hangers. Joists parallel to a wall should be strapped as shown in a previous drawing. Restraint strapping is unlikely to be found in properties built before the 1970s and all walls, particularly those parallel to joists, should be closely inspected for signs of movement. The consequences of failing to provide restraint can be seen in the chapter on Loadbearing Walls. If restraint has failed and the wall has moved away (ie outwards) from the floor there are usually some visual clues, such as:

- Bulges or bowing of the external wall.
- Cracks where ceilings join external walls.
- Gaps between the face of skirtings and the floorboards.

In older buildings of more than two storeys, the joists on alternate floors often run at 90 degrees to help restrain the walls.

Upper-floor defects

If joists are built in to a solid wall there is always the risk of dampness penetrating the end grain. This is less likely to occur if walls are over 1 brick thick. Even in cavity walls there is the risk of damp penetration if the joists project into the cavity, thus allowing the build up of mortar droppings etc.

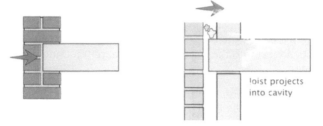

Joist projects into cavity

If the joist hangers are the wrong size, if the joists are too short, or if the joists are not nailed properly to the hangers, the floor joists could collapse.

If wall parallel to joists is not restrained the wall can move outwards.

These joists are built into a 1 brick wall. There is no restraint for the back wall (parallel to the joists).

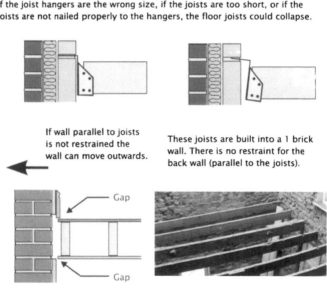

Gap

Gap

Sagging floors

One of the most common causes of deflection and 'bounce' is caused by failure of the strutting. If the strutting fails (or was not fitted in the first place), the floor is more likely to 'bounce'. The strutting should run at right angles to the joists (usually one line of struts in the middle of an average-sized room), with wedges or timber blocking between the outer joist and the external wall. This effectively forms a lattice beam at right angles to the joists. It stiffens the floor, limits 'bounce' and minimises twisting of the joists. Over time, minor building movement and timber shrinkage can loosen the wedges and blocking thus weakening the floor. If a floor 'bounces' when walked over, it is one of the first things to check.

Sagging floors can occur either through undersized joists or excessive loading, the latter often caused by change of use. Victorian and Georgian houses often had reduced joist sizes in bedrooms where high loading was not anticipated. Where these older properties are converted into flats the joists are often of insufficient size to cope with the increased loads of people, furniture and fire protection. Modern Building Regulations ensure that converted properties conform to current requirements, but the Regulations are not retrospective and thousands of properties were altered when legislation was less onerous. Repair can be very costly as it often requires the use of steel beams to support the floor mid-span.

In the late 1940s and 1950s there was a shortage of timber due to government restrictions on imports. In some houses built during this period the joist centres were 'stretched'; 600mm rather than 400mm centres was not uncommon. However, the joist depth was not increased to compensate.

If a joist is adequately sized the load it is carrying is transferred safely into the wall. If the joist sags it will tend to push the wall outwards (see the drawing on the next page).

Upper-floor defects – continued.

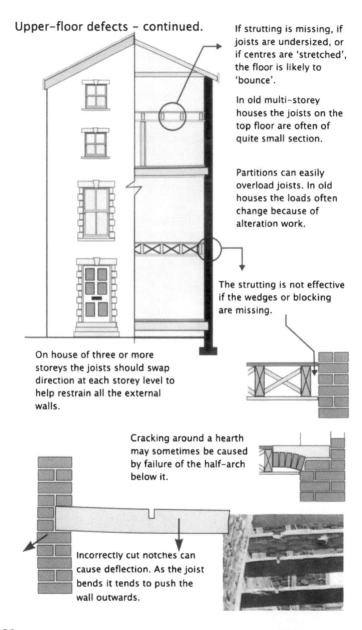

If strutting is missing, if joists are undersized, or if centres are 'stretched', the floor is likely to 'bounce'.

In old multi-storey houses the joists on the top floor are often of quite small section.

Partitions can easily overload joists. In old houses the loads often change because of alteration work.

The strutting is not effective if the wedges or blocking are missing.

On house of three or more storeys the joists should swap direction at each storey level to help restrain all the external walls.

Cracking around a hearth may sometimes be caused by failure of the half-arch below it.

Incorrectly cut notches can cause deflection. As the joist bends it tends to push the wall outwards.

Other problems with sagging may arise where supporting partitions have been removed during alteration works, or where joists have been excessively notched to carry services, such as central heating pipework and shower wastes. The drawing below shows the current rules for notches and holes. Do not expect, of course, old floors to conform.

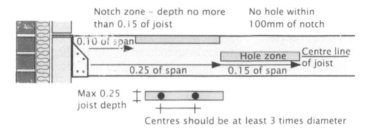

Notch zone – depth no more than 0.15 of joist

No hole within 100mm of notch

0.10 of span

Hole zone

Centre line of joist

0.25 of span   0.15 of span

Max 0.25 joist depth

Centres should be at least 3 times diameter

The photograph below shows how extensive notching has to be to accommodate central heating pipes.

Floor coverings

If a property is free from problems of damp penetration or insect attack, floorboards are rarely troublesome. However, other minor problems can occur, such as warping due to lack of nailing, or creaking due to the incorrect choice of nail. Chipboard is a durable material but can be damaged if it gets wet, eg due to baths overflowing. The problem is not caused by rot but by the chipboard swelling as it absorbs moisture. In addition, the moisture can break down the composition of the glue that binds the tiny timber chippings together. Moisture-resistant chipboard is available, designed for use in bathrooms and kitchens, but it will still fail if it remains saturated for long periods. See chapter on Ground Floors for more information on floorboards.

Cracked ceilings

In older properties cracking is often caused by the gradual deterioration of timber lath and plaster ceilings. Shrinkage of the laths, rusting of the nails that fix them to the joists, aggravated by vibration of the floor itself, will eventually lead to extensive cracking. Inadequate joist size and lack of strutting can, of course, exacerbate the problem. Plasterboard ceilings are less prone to cracking if properly fixed but, in practice, the following defects are common:

- inadequate nailing
- wrong type of nails
- poor taping of the plasterboard joints
- lack of timber noggings at unsupported edges
  of plasterboard.

These items are covered in more detail in the chapter on Finishes.

Another problem that can affect both modern and early floors is insect attack; this is also covered in more detail in the chapter on Rot and Insect Attack.

CHECKLIST (EXCLUDING FLATS)

Element	Comments/problems
From the rooms below...	Is there any evidence of ceiling damage? This could be cracking in the ceiling itself or cracking around its edges. Timber lath ceilings are prone to random 'map cracking'. Plasterboard tends to crack along its edges (joints).
From inside the room...	If the room feels or smells damp, rot should be suspected *(see later chapter)* and further investigation will be required.
Structural stability	Try the heel test – lift the heels and drop them quickly onto the floor. If the windows rattle or if furniture shakes badly there is a problem.
	Gaps exceeding 10mm or so between the skirtings and the boards may suggest movement (not just shrinkage) and require further investigation.
If you can lift the floor coverings to examine the floorboards...	Check the following: • Adequate thickness of boards. • Excessive gaps between floorboards. • Integrity of the boards – ie free from splitting, twisting and cupping. • Creaking, usually caused by incorrect nailing. • End joints (head joints) should be supported by joists. • Joist centres (measure gaps between nails in boards).
If the floorboards can be lifted...	Check the following: • Strutting and blocking at ends. • Integrity of the joists, particularly near the supports. • Joist size. • Damage caused by splitting, alteration work, building services installations. • Rot and insect attack. • Evidence of large notches and holes. • Are joist hangers tight against the wall and free from rust?

ROOF STRUCTURE

TRADITIONAL ROOFS

A simple roof comprises a series of sloping timbers (rafters) fixed to a ridge board at the top and a wall plate at the bottom. The wall plate sits on the top of the brickwork and, in traditional construction, is not mechanically fixed in any way. The roof stays in position because, when covered, it weighs several tonnes. To prevent the feet of the rafters pushing outwards they are securely nailed to the ceiling joists. A simple closed-couple roof (as this form is known) can span about 5m.

The ceiling joists, which span from plate to plate, are often of smaller section than rafters, and to prevent them sagging in the middle some form of intermediate support is required. If no intermediate support is available in the form of internal loadbearing walls, the most common solution is to secure the ceiling joists with a horizontal member, called a binder, and possibly to hang the binder from the ridge.

In a collar roof the tie is not at the feet of the rafter but some way up. A house with a collar roof has rooms formed partly in the roof space resulting in a lower overall height and subsequent savings in walling materials. However, if the collar is too high it will not be effective as a tie because the rafters below the collar can still exert a lateral thrust on the supporting walls. If the collar is kept fairly close to the rafter feet (not more than one-third up the height of the roof) and a good joint is formed with the rafters by either bolting or carpentry joints, the roof should remain stable.

At the bottom of the rafters, a timber fascia board protects the rafter feet from the elements and provides a fixing for the guttering. The fascia board can be flush against the brickwork or can provide an overhang. In the latter example, it is also common to find a soffit board supported by a simple timber framework nailed to the rafters. The soffit board prevents insects and birds from gaining access to the roof space. Overhanging eaves provide better weather protection to the wall below but are more expensive.

Simple pitched roofs

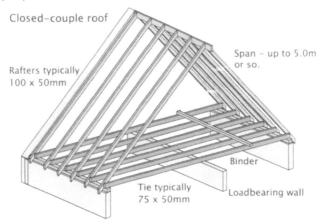

Closed–couple roof

Rafters typically
100 x 50mm

Span – up to 5.0m
or so.

Binder

Tie typically
75 x 50mm

Loadbearing wall

The ties provides support for the ceiling and prevent the feet of the rafter from spreading. In order to function properly the connection between the tie and the rafter must be sound.

If the tie is raised slightly there are minor savings in walling because part of the room is formed within the roof space. In this situation the tie is known as a collar. The integrity of the joint is vital.

Collar roof

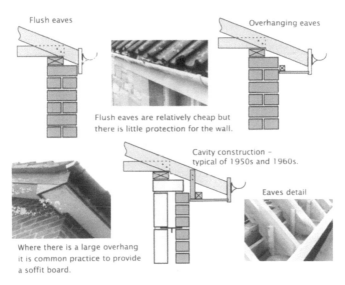

Flush eaves

Overhanging eaves

Flush eaves are relatively cheap but
there is little protection for the wall.

Cavity construction –
typical of 1950s and 1960s.

Eaves detail

Where there is a large overhang
it is common practice to provide
a soffit board.

Most houses with traditional or 'cut' roofs are too deep for simple
closed-couple or collar roofs. Houses two rooms deep normally
require roofs that can span at least 7m. The load from the roof
coverings, snow and wind will cause the rafters to deflect and will
place extra strain on the nails securing the rafter feet. In theory it is
possible to increase the span of a roof and prevent deflection by
using ever-increasing rafter sizes. However, the use of very deep
sections is both uneconomic and creates handling problems on site
due to their weight. In traditional construction, this problem was

solved by the introduction of a purlin. A
purlin is a large timber beam that
generally supports the rafters mid-span
thus preventing deflection. The purlin
spans from gable to gable (sometimes
with intermediate support) and besides
supporting the rafters provides lateral
stability to the gable end walls. The
purlins can just be seen in the bottom of
the photo.

It is a very common type of construction, particularly in its simplest form, where it is used extensively in terraced housing. As in an earlier example, some form of support will be required for the ceiling joists. This can be in the form of hangers, although it is more likely to be in the form of an internal loadbearing wall, which divides the front and rear bedrooms. As with timber floors, the ceiling joists are often in two pieces because long lengths of timber were not (and are not) readily available. Simple purlin roofs is suitable for spans up to 8m or so.

More complex purlin roofs

Most terraced housing is fairly narrow and purlins can easily span from gable to gable. However, as the width of a property increases so must the depth of the purlin if it is to be strong enough to take the load imposed on it without undue deflection. Very deep purlins, say over 250mm, are expensive, difficult to obtain and difficult to handle on site. In practice it is more economic to use smaller purlins and provide intermediate support in the form of struts, built off internal loadbearing walls. In a large terraced house or average-sized detached property it is common to find one or two pairs of struts, depending on the availability of supporting walls.

In modern housing steel purlins can sometimes be found. They can

span long distances and do not require as much intermediate support as their timber counterparts. They are usually supported on padstones to help spread the loads safely into the walls.

In some cases, load-bearing walls are not in the right position or do not exist at all and in these situations the purlins can be supported by trusses. Trusses can be found in various designs. Some are fairly simple; others are quite ornate. See page 106 for other examples.

Purlin roof

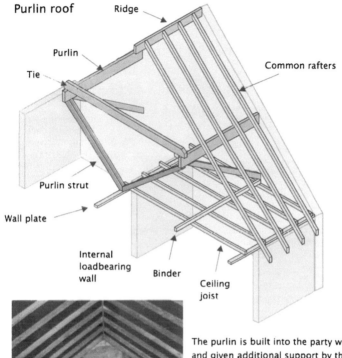

Ridge
Purlin
Tie
Common rafters
Purlin strut
Wall plate
Internal loadbearing wall
Binder
Ceiling joist

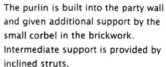

Tie

The purlin is built into the party wall and given additional support by the small corbel in the brickwork. Intermediate support is provided by inclined struts.

Purlin
Corbel

Tie
Purlin
Strut
Strut

105

Trusses

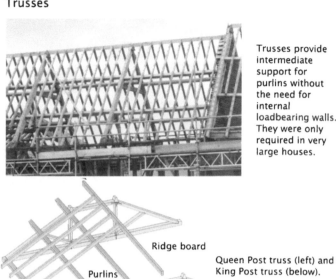

Trusses provide intermediate support for purlins without the need for internal loadbearing walls. They were only required in very large houses.

Ridge board

Purlins

Queen Post truss (left) and King Post truss (below).

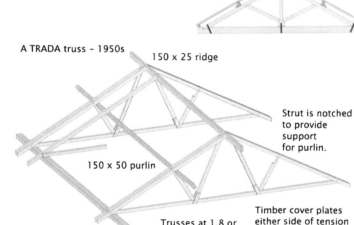

A TRADA truss – 1950s

150 x 25 ridge

150 x 50 purlin

Strut is notched to provide support for purlin.

Trusses at 1.8 or 2.4 centres.

Timber cover plates either side of tension members.

TRUSSED RAFTERS

Most modern roofs are constructed from trussed rafters. Trussed rafters have been popular for 40 years or so and have almost totally replaced the traditional roofs shown earlier. The most common pattern is the Fink or 'W' truss designed for symmetrical double-pitch roofs although there are a variety of shapes suitable for most roof designs. The trussed rafters are prefabricated and delivered to site ready for lifting on to the supporting walls, although occasionally you will find the entire roof structure assembled on the ground and lifted into place by a crane.

The timbers, which are typically 80 x 40mm in section, are butt jointed and held together by special plates, which are pressed into position by machine. The truss suffers minimal deflection under load due to the triangulation of the timbers and the metal plates that fix the members firmly together.

Trussed rafters offer several advantages when compared with traditional roofing methods.

- No internal support is required from loadbearing partitions.
- Spans of up to 12m can be easily achieved.
- They offer very fast construction.
- Skilled labour is not required.
- They are relatively cheap.
- They can be designed to very shallow pitches.

Perhaps their major disadvantage is that the use of the roof space for storage is severely limited due to the nature of the timbers.

Design and fixing

As with traditional roofs, a timber wall plate is bedded on the wall; the trussed rafters are fixed by nailing or by the use of special clips. Although they can be skew-nailed, through the holes in the plates, this can cause damage if not done with care and most manufacturers recommend the use of truss clips. Trussed rafters are usually designed to be fixed at 600mm centres.

Once in position, the trussed rafters must be linked, together with additional timbers, to form a single structural unit. These timbers also hold the trussed rafters in the correct position at the correct spacing. This is achieved by the use of binders (sometimes referred to as braces), which are nailed at right-angles to the trusses as shown in the diagram on page 109.

Although this binding ties the trussed rafters together, it does not stop the top part of the roof structure from deformation, caused by lateral loads, such as wind pressure on the gable ends. To prevent this, a diagonal brace is nailed to the underside of the top member as shown opposite. This diagonal brace prevents lateral deformation and keeps the structure rigid. Omitting the brace is a common cause of roof failure and this is further explained in the section on roof defects at the end of the chapter.

Strapping

Modern Building Regulations demand the use of straps in new roofs and it is a current requirement that the gable walls are tied back to the roof structure at intervals of not more than 2m. The strapping is to prevent movement of the gable walls under wind load. This strapping is required at rafter level and, in steep roofs, at ceiling-joist level to prevent the middle of the wall from buckling or bulging. The principle is exactly the same as for suspended floors, which were described in the chapter on Upper Floors.

It is good practice to strap the wall plate down to the external walls. This can be done using galvanised or stainless steel straps, which are screwed to the top of the wall plate and to the internal skin of blockwork. According to the Building Regulations, this form of strapping is not necessary for most roof structures but it is cheap to install and prevents the roof lifting in high winds.

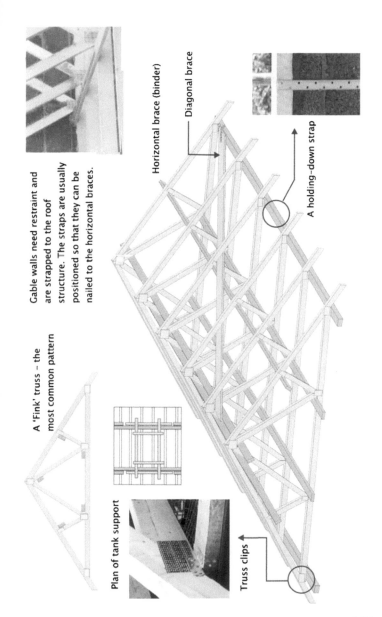

Gable walls need restraint and are strapped to the roof structure. The straps are usually positioned so that they can be nailed to the horizontal braces.

Horizontal brace (binder)

Diagonal brace

A holding-down strap

A 'Fink' truss – the most common pattern

Plan of tank support

Truss clips

GENERAL DEFECTS

A large proportion of roofing problems are caused not by defects in the design or construction of the structure but by subsequent problems of condensation, damp penetration and insect attack. Condensation is most likely to occur in old roofs that have been insulated without paying due attention to the provision of suitable ventilation. However, it can occur in new roofs that the insulation has been placed right into the eaves, thus blocking the flow of ventilating air.

Damp penetration is most likely to be a result of failure of the coverings. If the timbers remain wet for any length of time, dry or wet rot can occur and, in addition, nails and nail plates can corrode. Particularly vulnerable are those timbers that touch or are built into the gable ends. If a purlin rots it can affect the stability of the whole roof and, of course, the gable end which it is retaining. Fascia boards are also vulnerable areas. Overflowing gutters and inadequate painting can lead to wet or dry rot, which reduces the strength of the fascia, resulting in it eventually breaking away from the feet of the rafter.

Insect attack is always a risk in untreated timbers and this is explained in more detail in the chapter on Rot and Insect Attack.

Specific problems with traditional roofs

Roof spread

The roof can spread if there are no ties at ceiling level. In some cases the wall plates will be pushed outwards, but in more serious cases

the walls themselves will be subject to lateral thrust. Where ties exist they may be ineffective due to rusty nails. The problem can be exacerbated by re-covering roofs with heavier materials and poor original construction at the eaves.

Some roof shapes are unstable by their very nature and a good example of this is the double 'lean-to' roof. It is impossible with this shape of building to tie the rafters together and movement of flank walls is a common problem. The photograph shows a flank wall that is nearly 150mm out of plumb.

Undersized timbers

If timbers are undersized or placed at the incorrect centres, the roof structure can sag. If the deflection is serious it can dislodge the roof tiles and flashings, leading to problems of damp penetration. It can also place additional strain on the nails at eaves level. This is commonly caused by undersized rafters and purlins or ineffective strutting. If the roof undulates along its length, the problem is likely to be caused by rotten or undersized tile battens. In most cases mentioned above, additional timbers can be introduced to strengthen the roof but, in extreme cases, it may be necessary to strip the roof first.

Sagging ceilings

These can occur through ineffective hangers and binders, undersized ceiling joists or excessive weight of water tanks or materials stored in the roof space. In some cases the problem may have been caused by positioning and nailing the hangers and binders before the roof was covered. As the roof deflects slightly under the load the hanger actually pushes the ceiling down.

There are situations where sagging rafters have been strutted directly from the ceiling joists in an attempt to give them extra support. This is rarely effective and usually results in damage to the ceiling.

Settlement of internal walls

Some old houses have internal loadbearing walls built up to roof level to provide support for the purlin struts. If these walls have inadequate foundations, the load from the roof can force the wall downwards. This can often be identified by placing a spirit level on the joists and by inspecting the upstairs door openings, which are often out of square. If the wall is not central, the uneven thrust from the struts can also push the wall sideways.

Party walls

Party walls provide sound insulation, protection against the spread of fire and, in addition, often support the purlins. In some older terraces the party walls terminate at ceiling level, purlin support being provided in the form of columns of brickwork built up into the roof space. To prevent the spread of fire this gap should be filled with brickwork or timber studding covered both sides  with plasterboard. In the photograph, the party wall is not finished. In many semi-detached and terraced houses there is also the danger that fire can pass from one dwelling to another via boxed eaves.

Defects in trussed rafters

A variety of defects can occur in trussed rafters, most of them caused by poor site practice. One of the most common problems is inadequate bracing and binding. If these timbers are not correctly fixed, the roof structure will not be rigid and the rafters can develop a pronounced lean. This lateral thrust can push over the gable end walls and cause severe structural damage.

Even if the bracing and binding is correctly fixed, the gable end walls may still move as a result of wind pressure if they are not securely strapped to the roof structure. The strapping must work both in compression and tension to ensure that the gable end walls remain vertical. It is also good practice to strap the wall plates down to prevent uplift in high winds.

If the trussed rafters are handled incorrectly on site, the nail plates can be pulled out of the timbers with serious implications for the strength of the finished structure. When lifting the rafters into position, it is important that they are kept as near vertical as possible to avoid unnecessary strain on the plates. As mentioned earlier, the plates can also be affected by condensation, which, if unchecked over a period of time, can cause corrosion. Plate corrosion has also been identified as a result of spraying the roof space with insecticide; however, most modern treatments are quite safe in this respect.

Trussed rafters will deflect slightly when the tiles are laid and the erection of first-floor partitions should be delayed until the roof is complete to avoid unsightly cracking.

Finally, it is important to ensure that the trussed rafters are fixed at the correct centres. If the centres are 'stretched' the structure will be unstable and, in addition, the roof covering will undulate due to deflection of the comparatively thin tile battens.

VENTILATION AND CONDENSATION

If warm moist air passes through the ceiling and into the roof space it will condense on the cold timbers and roofing felt (under the tiles). The roof space is cold, of course, because of the insulation provided just above ceiling level, which retains the heat in the rooms below. Over a period of time this continual wetting of the timbers can lead to rot. In addition, ferrous nails and the nail plates can corrode. Condensation can be prevented by the use of roof ventilation to remove the moist air and by providing a vapour check immediately below the insulation. With ever-increasing levels of insulation, condensation has become a major cause of roof failure and is particularly common in older buildings where insulation is installed without provision of suitable ventilation.

Ventilation can be provided through the soffit board or above the fascia board. In this example the roof includes vents in the soffit board and a plastic tray to ensure that ventilation is not blocked by the roof insulation.

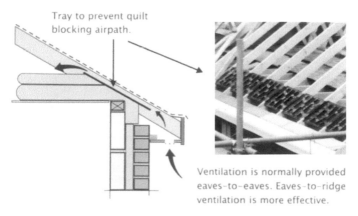

Tray to prevent quilt blocking airpath.

Ventilation is normally provided eaves-to-eaves. Eaves-to-ridge ventilation is more effective.

For normal double-pitch roofs (traditional and modern), there should be ventilation openings at eaves level to produce cross-ventilation. The most common way of providing this ventilation is by the use of specially manufactured soffit boards, plastic ventilating strips, or by leaving a gap between the soffit and the wall. Some tile manufacturers also produce plastic vents that fit under the bottom course of tiles. The ventilation should normally be equivalent in area to a continuous 10mm gap along each side of the roof.

In shallower roofs (below 15 degrees), which contain a smaller volume of air and are therefore most likely to suffer from saturation, the gap should be increased to 25mm.

Eaves-to-eaves ventilation is recommended by the Building Regulations although it is not always completely effective. Ventilation only occurs when the wind is blowing and there is the risk of stagnant warm moist air being retained in the upper part of the pitch where, evidence suggests, most of the condensation occurs. A better form of ventilation is to use eaves-to-ridge ventilation. This has the advantage of ventilating the whole of the roof void and also functions in conditions of no wind, due to the natural convection currents that occur in the roof space.

On no account should ridge ventilation alone be used. As the wind blows over the roof, it creates a vacuum in the roof space. This will suck air from the rooms below, increasing the transfer of moisture vapour from the building into the roof space.

If there are rooms in the roof space, the provision of ventilation is more complicated, as part of the insulation will follow the line of the rafters. To ensure a flow of air, in this situation there should be a gap of at least 50mm between the top of the insulation and the roofing felt. Failure to provide this will result in condensation on the roofing felt, wet insulation, stained ceilings and potential problems of rot. This is a common cause of failure, particularly in existing roofs that are subsequently converted into rooms.

Vapour checks

The use of vapour checks (vapour control layers) at ceiling level will help to reduce the amount of water vapour that finds its way into the roof space. These can be in the form of polythene laid under the insulation, or foil-backed plasterboard. However, due to the joints in the vapour check and holes cut for loft hatches and ceiling lights, they should not be regarded as completely reliable and should be considered in addition to ventilation, not in place of it.

Breathing roofs

In recent years the provision of breathing roofs has become more common. In a breathing roof there is no eaves or ridge ventilation. A vapour check must be provided to minimise the amount of moisture entering the roof. Unfortunately, breathing roofs are not fully understood by all builders. The use of a vapour permeable roofing felt does not, in itself, always preclude the need for ventilation. See the checklist for more details

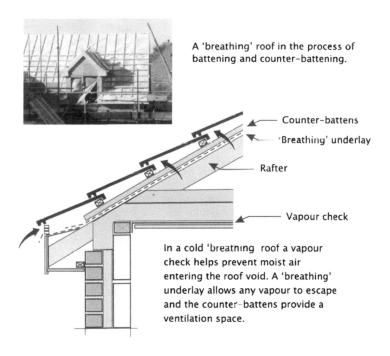

A 'breathing' roof in the process of battening and counter-battening.

Counter-battens

'Breathing' underlay

Rafter

Vapour check

In a cold 'breathing' roof a vapour check helps prevent moist air entering the roof void. A 'breathing' underlay allows any vapour to escape and the counter-battens provide a ventilation space.

CHECKLISTS

We have divided this section into three parts:

- traditional roofs
- trussed rafters
- ventilation, fire etc.

TRADITIONAL OR 'CUT' ROOFS

Element	Comments/problems
Evidence of dishing	From outside the property examine the roof slopes – are they free of sagging or dishing? Are there any signs of spreading?
Sagging ceilings	Before getting into the roof space have a look at the upstairs ceilings. If they are sagging, it could be a plaster problem but it could also be a problem with inadequate ceiling joists, or a lack of, or problems with, hangers and binders. A roof containing hangers can push a ceiling downwards if the roof is recovered with heavier materials.
Rafters and ceiling joists	From within the roof space, check the condition and spacing of the roof timbers. Timbers in existing roofs will not be as substantial as those required by current Building Regulations. However, if they are free from sagging or deflection, they are obviously doing their job. The joint between the rafters and joists should be sound if roof spread is to be avoided. This is particularly important in a collar roof.
	Always be alert to rot and insect attack (see chapter on Rot and Insect Attack). In the case of rot remember that it is the roof perimeter (verges, ridge and eaves) that is most likely to suffer from saturation and therefore rotten timbers.
Hangers and binders	Ceilings may sag if hangers and binders have broken free (nails rust). They may also have been cut out to clear the loft space.

Element	Comments/problems
Purlins	As with rafters most existing purlins will be undersized according to current Building Regulations. However, if they do not sag they should not be a concern. Check the purlin ends for dampness – they are vulnerable where they are built into gable walls. Also check the integrity of the support for the ends of the purlins.
	Purlin struts normally bear on an internal loadbearing wall. Check their junction with the wall and examine the wall itself (from within the house) for signs of settlement. Doorways out-of-square suggest settlement. This is a particular problem where roofs have been re-covered with heavy materials.
	Some purlins may be supported by heavy timber trusses. Check the truss itself for splits in the timber and the integrity of the joints. If possible check the supporting walls for signs of overloading.
Cisterns	Check that cisterns have adequate support. See trussed rafters below for more detail regarding support boarding.

TRUSSED RAFTERS

Trussed rafters	Centres should normally be 600mm. They should be evenly spaced and vertical. They should normally be doubled-up where the roof is trimmed, eg either side of central stack. None of the truss members should be missing and none should be split. New trusses should have visible stress grading and quality assurance marks on the timbers.
Plates	Check that the nail plates are securely bedded in the timber and check that there no signs of premature corrosion. The trusses should be fixed with truss clips, or with nails through the plates (no longer recommended). In either case check the nailing is adequate. The plate itself should be bedded in mortar and should normally sit on the internal leaf of a cavity wall.

Element	Comments/problems
Bracing	Check that there is longitudinal bracing at each node point (normally five braces are required) and that they are lapped where they are not continuous. The braces should be butted against gable walls. Diagonal braces should be fixed either side of the slope and should be nailed to the plate at the bottom. Note that large span roofs (over 8m or so) may require additional bracing.
Strapping	Wall plates are normally strapped down at 2m centres. The need for straps depends on roof pitch, exposure, weight of coverings and local conditions. Gable restraint straps are nowadays required at the verges (at 2m centres). The straps should extend across three trusses, with timber blocking in between. On steep roofs straps should also be provide at ceiling level to restrain the centre of the gable wall. Remember that in older roofs (mostly from the 1970s and early 1980s) requirements for strapping and bracing were not as onerous as they are today.
Cisterns	They should normally be supported on a series of bearers spanning at least three trusses. Insulation should be over, and up the sides of cisterns, but not underneath them. Bearers should be close to node points of trusses. Tank platforms should not be made from chipboard as it is too easily damaged through condensation. In extreme cases the tank can fall through the ceiling.

ALL ROOFS – FIRE

Fire precautions	Check that party walls run right up to the roof-line and that there is adequate fire stopping over the party wall (not always that easy to check in practice). Fire stopping should also fill any boxed eaves. There should also be fire stopping in the cavity at the top of a cavity wall (just below plate level). This can be in the form of cavity batts (filling the cavity), a cavity barrier (for open or partial fill cavities) or a cavity

Element	Comments/problems
	closed in blocks or bricks. Timbers should not be built into the party wall (although purlins usually are – if so, the brickwork must be a good fit around the purlin).

ALL ROOFS – VENTILATION AND INSULATION

Breathing or ventilated roof	The vast majority of roofs are traditionally ventilated. There should be an air gap along the eaves and the draft should be felt within the roof void. If the roof is a complex shape check there are no stagnant areas. In a 'breathing' roof there should be a vapour check at ceiling level, seals around services running through the ceiling, permeable felt and counter-battening. If the last point does not exist, adequate ventilation may still occur as long as the felt sags in between the rafters.
Is the roof felted?	Roofs are felted to prevent snow and dust being blown into the roof space. The felt also reduces the wind load under the tiles. If the roof is not felted It is not necessarily defective. Where it does exist it should not be stretched tight over the rafters and it should be free from tears. The felt should have adequate laps. See next chapter on Tiling and Slating for more information on underfelt.
Moisture under felt	If the underside of the felt feels wet it is most likely to be a result of condensation.
Insulation	Check the depth of insulation – nowadays it should be 200–250mm. The insulation should be free from gaps and should be laid into the eaves – but eaves ventilation channels should not be blocked. Where possible, insulation should not be laid over electrical cables as they may overheat.
Services	All water and overflow pipes should be protected against frost.

TILING AND SLATING

MATERIALS

In modern construction most roofs are covered with concrete tiles. However, natural slate, artificial slate, clay tiles and even thatch are still common.

Thatch has been used in rural areas for centuries; it was also common in towns until the end of the medieval period. The main thatching materials are water reed and wheat straw. The life of thatched roof can vary from 15 to 50, years depending on the material, pitch, thickness, and exposure.

Natural slate is a hard-wearing, traditional roof covering, which can easily last 200 years. Some 100 years ago, most slate was quarried in Wales – much is now

imported. The quality of slate varies from quarry to quarry and this is reflected in its price. Many speculative Edwardian and late Victorian houses were built with slates of average (or less) quality; these inferior slates tend to be very dark, almost black.

Stone slate, unlike natural slate, is often made from limestone or sandstone. It is generally not as durable as natural slate. Stone slates vary from region to region depending on the local geology. Synthetic or artificial slates have been available for almost 100 years. They were originally made from asbestos cement and look a bit like natural slate but are much lighter and cheaper. Some artificial slates (right) are made from concrete and are reproductions of traditional stone slates. In recent years the quality of artificial slates has improved immensely.

Plain tiles have been used in this country for centuries, particularly in the 'lowlands' where natural clay was in plentiful supply. Plain tiles are still made from clay and are, once again, becoming popular. Concrete plain tiles have been available since the 1940s.

Double-lap tiles and slates

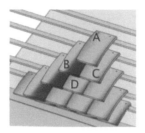

Slates and plain tiles are double-lap materials. Because they do not overlap at the sides, an extra layer of tiles or slates is required to prevent water getting into the roof space. If water runs off tile 'A', some of it will penetrate the joint between 'B' and 'C'. This is fine because another tile, 'D', is there to catch it.

These are plain tiles. They can be made from concrete or clay. The tiles have small nibs on the back which fit over the battens. It is normal practice to nail the perimeter tiles and every fifth course. The minimum pitch is usually about 40 degrees, the minimum lap is about 65mm (the amount by which 'A' overlaps 'D'). They weigh about 75kg/m^2.

Slates vary in size and thickness. Because they do not have nibs every slate has to be nailed to the battens. The minimum pitch depends on the size of the slates; slate 600 x 300mm can be laid to a pitch of about 25 degrees as long as the headlap is 100mm or so. They weigh about 30–40kg/m^2 (Welsh slate).

Slate can be head nailed or centre nailed. If the latter is used it is important that the roof is designed (slate width and pitch) so that water running off 'A' and into the gap between 'B' and 'C' cannot penetrate the nail holes on slate 'D'.

Slates should normally be supported by 25 x 50mm softwood battens. Deep battens (at least 25mm) are important if 'springing' is to be avoided – springing prevents the nail from being driven home completely – the batten just flexes.

Single-lap tiles

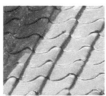

Pantiles

Single-lap tiles account for the vast majority of tiles sold today. With these tiles there is only one layer of tiles on the roof - apart from the overlaps of course. Early single-lap tiles overlapped at the side, eg pantiles and romans. For the past 80 years or so they have been made with more sophisticated side joints (interlocks). Water cannot penetrate the side joints due to the overlap or interlock on the tiles.

Single and double romans
These overlap at the sides.

Modern profiled concrete tiles

These (left and right) interlock.

Not all interlocking tiles are profiled. Some are flat. Some are even made from crushed slate - they look like natural slate but they interlock and are therefore single lap.

Pitch - as low as 15 degrees (depending on tile)
Weight - about 45kg/m^2 (concrete and clay)

Single-lap flat interlocking tiles (left), and profiled (below).

As a general rule, the perimeter tiles should be fixed by nailing, or by the use of special clips, and on steep roofs, say over 45 degrees, it may be necessary to clip and nail all the tiles. In practice, some modern tiles have to be fixed by clips as they do not have nail holes.

Pantiles were introduced to this country some 300 years ago. They are single-lap tiles (unlike slates and plain tiles) and early examples can be found in the South-east and South-west. Single and double romans were developed from pantiles. They offer slightly better weather protection than pantiles because of their closer fit. They were popular until the 1940s but have been superseded by interlocking tiles.

The vast majority of modern roof coverings are concrete interlocking tiles. They are available in a huge range of colours and shapes.

Underfelt

Modern roofs have a layer of underfelt just below the tiles. Felt was introduced in the middle of the 20th century. Then its primary function was to reduce the flow of air through the roof, thus preventing the ingress of wind-blown snow or rain, and improving the roof's insulation. Nowadays felts also help reduce wind loading (wind pressure under the tiles) and act as a second line of defence in case of leaks. If any rain or snow does penetrate the tiling it must be free to escape and, therefore, it is important that the felt is correctly laid to enable the water to run down the roof slope and into the gutter.

Today felts can be made from a variety of materials, some of these are microporous. Water cannot penetrate these microporous felts but vapour can escape, thus helping reduce the risk of condensation. The need for and nature of roof ventilation was explained in the previous chapter. For more general information, including potential pitfalls, see the drawings on page 125.

Flashings

Flashings are required whenever the tiling or slating abuts a vertical upstand. In the past, flashings have been made from lead, copper, zinc and even mortar. Nowadays lead is generally the preferred material. It is durable, easy to cut and easy to shape. Where valleys occur (where two roof slopes run into each other) the valley lining can be formed using special tiles, lead sheet and, more recently, glass fibre valley sections. Long lengths of lead are unable to endure the effects of continual thermal movement. Splitting or bucking is almost inevitable. For this reason, the length of lead sheet used in flashings or valleys needs to be limited. The maximum permitted length depends on the pitch, the thickness of the lead and the methods of fixing.

Roofing underlays (felts)

Rolls of roofing felt are usually 1m wide and can vary in length from 10m to 50m. The felt is laid horizontally across the roof starting from the eaves and should be lapped about 150mm both vertically and horizontally. It should sag slightly between the rafters to ensure that any water finding its way through the tiles can drain down the felt to the gutter.

At the eaves felt should be supported by a timber fillet (to prevent sagging behind the fascia) and, ideally, there should be short strip of external grade felt at the eaves.

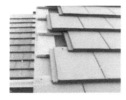

Felt is normally held in place by the tiling battens – these will vary in size depending on the type of covering and the rafter centres, 38 x 19mm, 38 x 25mm, and 25 x 50mm are common. At a verge, a composition board undercloak normally sits over the felt; this supports the verge pointing.

Some roofing felts are quite brittle and are easily torn. They are most vulnerable where exposed at the eaves; hence the reason for fitting a better-quality felt strip at eaves level.

Many older roofs have been re-felted using modern microporous underlays. In certain situations, these may not require roof ventilation. Ideally, these roofs will have a vapour control layer at ceiling level and counter-battens (to help ventilation above the felt). Always look carefully for signs of condensation if you cannot feel fresh air blowing through the roof void.

Flashings

Mortar flashings are less effective than metal ones due to their inability to accommodate thermal and structural movement, and the risk of frost attack (both result in cracking). In this example, cantilevered slates are directing water away from the roof/wall joint to provide additional protection.

Lead is durable, easy to work, and is available in a range of thicknesses. In the first example (below) the flashing comprises a number of soakers with a cover flashing over the top. This can be used for flat or profiled tiles. An alternative approach is to use a series of one-piece abutment flashings.

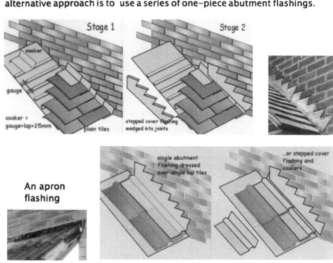

An apron flashing

Flashings can fail for a variety of reasons. The most common include, wind damage, no provision for thermal movement (lead splits or buckles), over 'dressing' (lead becomes thin), inadequate fixing and wedging, inadequate size (usually the width of soakers), and incorrect grade of lead.

This stack flashing has been patched with a thin lead-backed adhesive tape. The valley on the right should have been laid in shorter lengths (with steps at the junctions).

DEFECTS

Considering their exposure, most tiled and slated roofs are surprisingly durable, and a life of 100 years is quite normal. However, a few common problems can occur; some related to the materials themselves and others related to poor design or bad workmanship.

Design

The roof coverings should be appropriate to the pitch of the roof. Some of the modern interlocking tiles (and some wide slates) can safely be laid to very shallow pitches but plain tiles require at least 40 degrees to maintain weather protection.

The majority of roofs are designed to direct water away from the building itself, but there are some examples (mostly pre 1900) where rainwater is collected in a valley or lead-lined gutter and then channelled to a downpipe. These forms of construction may have been adequate when first built but, over the years, deterioration of the lead, together with minor structural movement of the roof timbers, can often lead to problems of damp penetration, particularly if snow is allowed to collect on the roof.

Lead gutters and valleys should be stepped at regular intervals to allow for thermal movement of the lead, but this is only common in more expensive properties and it is not unusual to find lengths of lead several metres long. Where this is the case, expansion of the lead can create a series of small ridges, which can ultimately lead to cracking and water penetration.

General problems

In many cases, the first part of a roof to deteriorate is not the tile or slate covering, but the battens underneath. These can fail due to rot or insect attack and can also become displaced due to rusting of the nails. If the battens are undersized the roof has a rippled appearance as the battens sag between the rafters. The felt underneath the battens is also an area that can fail prematurely. Early felts became quite brittle with age and tear easily.

Problems of damp penetration can occur if verge pointing becomes dislodged, if the ridge riles are blown out of position, or if the tiles themselves crack. Blocked gutters and downpipes can cause damp at eaves level, resulting in problems of damp penetration and the potential risk of rot in the fascia and rafter feet.

Metal flashings can corrode due to chemicals in the atmosphere and can tear if insufficient allowance is made for thermal movement. The mortar flashings that are sometimes used in their place are rarely successful. Thermal and moisture movement usually result in cracking of the flashings and damp penetration.

Slate

The quality of slate depends on its thickness as well as the quarry from which it was produced. The durability depends mainly on the slate's resistance to acid attack. Carbonates present in the slate react with the acid to form calcium sulfate, the crystals which force the thin laminations apart. Decay often starts underneath the slate where moisture has been drawn by capillary action and therefore the signs of decay are easy to miss during a building inspection.

Imitation slates are affected by moss and lichen, which attack the surface of the slates and cause slight softening. Atmospheric pollution, particularly acids, can cause surface softening and can also result in discolouration.

Stone slates are susceptible to frost attack. The slightly porous stone absorbs rainwater, which then expands in freezing weather and can result in cracking.

The fixing of all slates can be affected by the rusting of nails. Ferrous nails are no longer used (at least they shouldn't be) in tiling and slating.

Clay tiles

Towards the end of the last century there was a huge increase in the production of clay tiles and some early tiles often contained impurities, such as limestone and chalk, which, on firing, formed small nodules of unslaked lime. In addition, if the clay contains soluble salts the tile can be damaged as crystallisation occurs. Clay tiles are moulded in power presses and in these mechanical processes the tile can acquire a laminar structure. Fusion may occur as the tile is fired, but inadequate firing can result in future problems of frost attack occurring in between the laminate. This is most likely to occur where tiles have been laid to shallow falls, and where the tile remains wet for long periods.

Defects

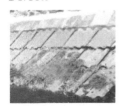

Above left – the tiles have been laid with too small a lap. Moss also slows down drying and impedes ventilation. The tiles on the right do not fit properly and have been crudely patched.

Above left – cracked verge pointing and notice adhesive repair at abutment. Centre – ridge needs repointing. Right – cracked tile (out of sight – behind a parapet – on a brand new house).

Above left – wind damage to pantiles. Centre – long-term frost damage on shallow pitched, moss-covered plain tile roof. Right – tingles securing slipped slates and moss growth on replacement artificial slates.

Above left – broken nib on plain tile. Centre – broken nibs on interlocking tile. Right – damage at verge on plain tile roof.

Concrete tiles

Some of the early tiles failed due to frost attack and sulfate attack. There were additional problems of discolouration and, in some cases, the rough surface of the tiles attracted dirt and resultant moss growth. However, modern concrete tiles are made to the highest quality and several manufacturers offer guarantees of 50 years to demonstrate their faith in their products.

Finally a note of caution. If an old slated roof is to be recovered with concrete tiles, special care should be taken as the roof structure may need strengthening to cope with the increased load.

CHECKLIST

Element	Comments/problems
Roof structure	Does the roof structure seem true, ie free from dishing or other movement? A badly dishing roof can dislodge the tiles and allow water to penetrate.
Tiling and slating	Inspecting from the ground is not always that easy. A neighbour's window and a pair of binoculars help. If it is a detailed inspection a ladder may be worth considering. Check for the following: • Cracked, slipped, dislodged or missing tiles. • Deterioration of tiles. • Evidence of past repair, eg tingles, lead-backed adhesive. • Tiles and slates with inadequate laps. • Moss and lichen.
Ridge and verge	The perimeter of the roof is its most vulnerable part. Check the ridge for missing or cracked tiles, signs of movement, and deterioration of pointing. Note that some ridges are dry-fixed (mechanically fixed). There are various fixing mechanisms, some more reliable than others. Check the verge for missing or cracked tiles, or broken verge clips. The verge infill pointing is also vulnerable. Timber barge boards (just below the verge) can deteriorate quickly if they are not painted regularly.

Element	Comments/problems
Eaves	This is a very vulnerable part of the roof because this is where all the water normally discharges. The tiles should overhang the inner edge of the gutter slightly. Felt should project just beyond the tiles and, preferably, should be an external grade (for durability). The felt should not sag behind the fascia board as this can lead to ponding. Check the condition of the soffit and fascia boards.
Lead work and flashings	Check for the following: • Signs of stress (tearing, buckling) etc, wind damage, evidence of theft. • Adequate covering of the tiles or slates. • Adequate lap with other flashings. • Adequate upstands, properly secured into the wall. • Provision for thermal movement (multiple sheets of lead rather than a single long length). • In lead valleys check that the leadwork is free to expand and contract and that there are properly formed steps where sheets overlap.
Guttering and downpipes	The best time to check guttering is when it is raining. Check the following: • Gutter size and fall. • Position – ie does it catch all the rainwater? • Condition of unions, stop ends, number of support brackets (is there any sagging?). • Evidence of leaks, or overflows down the wall. • Condition and fixing of downpipes. • Safe discharge (eg not onto footpaths).
From inside the roof...	If you can get into the roof space, have a good look at the felt (assuming it exists). It should not be stretched tight (unless the roof is counter-battened Check for tears – generally and around soil vent pipes. If the felt is wet on the underside, check for ventilation (see previous chapter). If the roof is not ventilated assess whether this is through design or error.
Insulation and ventilation	See previous chapter

CHIMNEY STACKS

INTRODUCTION

It is easy to ignore chimney stacks when inspecting a building. They do, however, give rise to a number of defects, including structural failure, dampness and, of course, leakage of gases and smoke. All these can be significant hazards. Remember, also, that chimney stacks that appear fairly sound from the ground may look much worse on closer inspection.

Shortage of space precludes a detailed explanation of the evolution of chimneys. For the same reason, we have had to omit any information relating to the operation of fires, eg the nature and condition of flues, fireplace, hearths and ventilation. If problems are suspected a heating engineer should be consulted. We have, therefore, concentrated on the physical nature of chimney stacks and some of the most common problems that affect them.

The following three pages of graphics show traditional chimney construction and include a number of common problems, some related to structure, others related to dampness and condensation. The inspection checklist includes some observations and comments on failure.

Note that many problems put down to penetrating dampness are, in fact, caused by condensation. Staining of plaster, for example, is a common problem on chimney breasts. It is often assumed that this is due to water running down the stack itself, or possibly due to flashing failure where the stack runs through the roof. Although this can be the cause of staining, there are other possibilities. Ammonia and sulphur dioxides (from solid fuel or gas or oil) form salt deposits within the stack or flue. These salts are hygroscopic and are often accompanied by rusty looking stains. Condensation (or rain water) occurring in a flue can absorb these soluble salts and deposit them on the plaster surface when the water eventually evaporates.

Typical late Victorian terraced house

In the roof space the four flues (two front rooms, two back rooms) are gathered to feed into one stack. The render limits smoke and fume leakage.

Chimney stack serves eight flues, four either side of the party wall. Each flue has its own chimney pot.

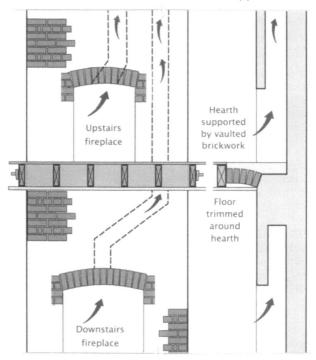

Upstairs fireplace

Hearth supported by vaulted brickwork

Floor trimmed around hearth

Downstairs fireplace

See floor plans in Ground Floors and Upper Floors chapters for more details of trimming.

Stacks – structural problems

Both these stacks have been affected by sulfate attack. See 'Walls' for a more detailed explanation. It can cause cracking and leaning – the latter because the rate of attack varies depending on saturation.

Sulfate attack cannot be repaired. Taking down and rebuilding is the only solution. It can be prevented by using sulfate-resisting cement.

Where chimney stacks are on hipped slopes they have to be quite high. Height should not exceed 4.5 times the narrowest width. Stacks not conforming should be removed, rebuilt or restrained. On tall stacks the quality of the mortar and pointing is particularly important.

From the ground stacks may look fine. Binoculars may reveal a number of problems:

- Loose or cracked pots.
- Cracked or loose flaunchings (the mortar holding the pots in place).
- Frost or sulfate attack.
- Loose pointing.
- Vertical splits in the stack – usually caused by expansion of linings or inner part of chimney.
- Excess loading or other damage from aerials.

Stacks – damp problems

High stacks, centrally positioned, were often built without DPCs. In a well ventilated roof void any moisture should evaporate.

Chimneys in this position (below) should include a horizontal DPC just above, or just below, the roof line. Some will have both (see photo below). A few will have a stepped DPC – this is an acceptable alternative.

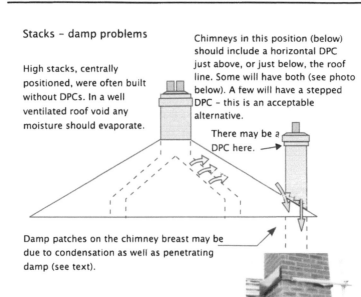

There may be a DPC here. →

Damp patches on the chimney breast may be due to condensation as well as penetrating damp (see text).

Flashings can be made from a variety of materials including lead, copper and zinc. The life of these materials depends on their thickness. Typical flashings are shown below.

Some chimneys have mortar flashings – treat these with suspicion. They are brittle and prone to cracking. In thatch roofs they are often unavoidable.

Back gutter with cover flashing.

Side flashing can be of various types.

Front apron flashing.

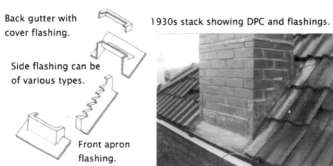

1930s stack showing DPC and flashings.

CHECKLIST

NB: stacks often appear to be in good condition from the ground. A thorough inspection may require ladders. At the very least, get some decent binoculars. Sometimes, a look from an adjoining house window can provide another useful chimney view.

Element	Comments/problems
Chimney height	Is the chimney too high for its width (see guidance on previous page)? Is the top of the chimney well clear of any skylight, ventilation duct or adjoining building? If the chimney is within 600mm of the ridge, it should be 600mm higher than the ridge. Beyond this, it should be at least as high as the ridge.
Pots and flaunching	Are the pots cracked or broken, do any appear to be missing? Is the flaunching in sound condition? Have old flues (ie not in use) been capped over without regard to stack ventilation?
Chimney stack	Assess the condition of the stack for: • Condition of the mortar joints. • Signs of sulphate attack. • Signs of leaning. • Cracks in the brickwork. • Frost attack in the bricks or stonework. • Potential damage caused by the weight (lever action) of aerials. Tall stacks (where height is greater than 4.5 times the width) that lean are potentially dangerous. Leaning stacks that are lower can sometimes be repointed.
Damp penetration	Check the following from outside the roof: • Nature and condition of flashings. • Position of DPCs (if they exist). Check the following from inside the roof: • Dampness on the chimney brickwork or on adjacent timbers, usually caused by defective flashings (note timbers should not be built into the chimney).

Element	Comments/problems
Damp penetration	Check: • The condition of the mortar pointing and the render (if it exists). • Make sure that any disused stacks that terminate within the roof void are adequately vented to the outside. Check the following by looking at the chimney breast below the roof: • Signs of dampness, staining or damaged plaster on the chimney breast. They could be caused by flashing failure, lack of DPCs in chimneys, rainwater running down the stack, and condensation. • Check to see whether disused flues are vented at hearth level.
Chimney operation	This book does not include information on the safe operation, combustion and exhaust of a fire. The chimney must be designed and built so that there is a good up-draught, complete combustion, and safe passage of the combustion gases to an appropriate point outside.

RENDERING

INTRODUCTION

Renders, in one form or another, have been used for hundreds of years to decorate or weatherproof buildings. During the Regency (early 1800s) rendering became a very popular finish (it was known as stucco) used, in the main, on cheap speculative brick buildings to emulate stone. Cheltenham and Brighton have hundreds of examples (the picture on the right shows a terraced house in Weymouth – late 18th century). Stucco was made from lime and fine aggregate. The type of lime varied but weak hydraulic limes were preferable to 'pure' limes. They had a faster set and better waterproofing qualities. Local aggregates would have been used; these might include sand, grit, and ash. During the next hundred years, building technologists experimented with various mixes of lime and clay, and produced a range of cements of varying strength and quality. The one most commonly used today is Ordinary Portland Cement.

A typical render from the early part of the 20th century would have been applied in two or three coats. It could have hydraulic lime as the binding agent or cement. If it was cement-based it would probably include a small proportion of pure lime to improve its working and weathering qualities. A typical mix would have been 1 part cement, 1 part lime, and 6–7 parts sand (or other fine material). Renders were usually smooth faced or rough cast. In the latter, a course aggregate is mixed in with the final coat. Do not be deceived into thinking stucco is actually fine stonework. Renders were often incised with fine lines to look like ashlar.

Modern renders

Apart from conservation work modern renders are nearly always cement-based although lime or chemical additives can be added to make the mix more workable and to improve its durability. On exposed buildings or where the wall surface is fairly rough, three coats will be necessary. In other situations, two will normally suffice. The top coat can have a smooth finish, a textured finish or can contain a coarse aggregate to give a variety of decorative effects.

General principles

Achieving a high-quality, durable finish depends on a number of criteria. These include:

The nature of the background. Good-quality bricks and dense blocks with raked out joints provide the best background for rendering. Smooth materials, such as insitu concrete, will need adhesives or special surface treatments to be successful. Old porous brickwork or smooth stonework may need covering with mesh.

The suction of the background. If suction is too high (some aerated concrete blocks for example), the render might dry out too quickly. Water required for hydration (the chemical set) will be sucked into the blocks.

The strength of the render. Strong renders should be avoided (eg 1 part cement: 3 parts sand). They are prone to shrinkage and cracking. Weaker mixes such as 1:1:6 (cement:lime:sand), are preferable. They can more readily cope with the extremes of thermal and moisture movement and, even though a weak render will absorb some rain, the relatively porous nature of the render allows evaporation to occur when the weather changes.

Thickness of coats. Trying to apply thick coats is rarely successful; it is impossible to squeeze all the air out from behind the render and get a good bond with the background. On a good surface a base coat of about 10–12mm is fine with a top coat of about 6mm.

The surface finish. Renders with a textured or aggregate finish provide better weather protection than smooth renders because the rough surface helps spread the rainwater evenly across the wall. On a smooth render it is often possible to see long streaks of water where excessive rain has been directed onto it through errors in design. This can lead to saturation and potential problems of damp penetration. Smooth renders also require more skill to achieve a good finish.

Render – general details

Three common render finishes: smooth, pebble dash and rough cast. Poor backgrounds such as stone and old bricks do not provide a good render key. Stainless steel mesh screwed to the background is the only effective solution.

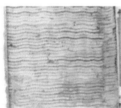

Once the undercoat is laid on the wall, it should be scratched to form a key for the next coat. The scratching also provides a series of stress relief lines to control shrinkage.

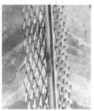

External angles are usually formed with angle beads. These should be stainless steel. Traditionally, the angles would be formed in render using vertical battens as formwork.

Design details are important in rendering. Rendering over DPCs, for example, may allow rising dampness to bridge the DPC. Similarly, existing buildings may need alterations to sills and parapets if drips are not to be blocked.

A bell mouth can be formed at DPC level or over windows to provide a drip.

DEFECTS

Shrinkage cracking

This, as its name implies, is caused by shrinkage of the render and commonly occurs where strong render mixes are used on weak, or badly prepared, backgrounds. If the cracks are only in the topcoat, the likely cause is excessive cement in the finish coat or excessive thickness. It can also occur if the render mix is too wet. Shrinkage cracking is usually easy to recognise by its distinctive pattern, as shown on the right.

Surface crazing

This is generally less serious than shrinkage cracking and most commonly occurs in smooth renders that have been over-trowelled. Excessive trowelling brings a thin film of water/cement paste to the surface, which forms tiny hairline cracks as it dries. It is sometimes only noticeable after rain.

Loss of bond

If the wall sounds hollow when tapped, it is likely that the render has not bonded properly to the wall. This can be a result of shrinkage cracking in a strong render, but also is caused by poor mechanical key and inadequate suction. If the problem occurs after a spell of very

cold weather, it may be caused by frost attack. If water trapped behind a dense render turns to ice, the associated expansion can force the render away from the wall.

Damp penetration

This can occur if very weak renders are used in exposed situations, but is more commonly a result of shrinkage cracking caused by rich mixes. Once the water has penetrated the cracks, the rich dense mix prevents subsequent evaporation. Rising dampness can also occur where the render has bridged the DPC. In 1920s and 1930s houses it is quite common to find render running down to ground level and bridging the DPC. In some cases, differential movement above and

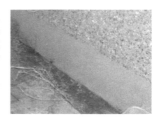

below the DPC causes horizontal cracking that allows rainwater, running down the render, to enter the wall.

Chemical attack

The most serious example of this is sulfate attack. If a strong render is applied to the wall, and it subsequently shrinks, rain can penetrate the cracks. If the wall behind the render becomes saturated, soluble salts, which are present in some bricks, can react with the cement in the mortar to produce a chemical compound that rapidly expands as it forms. This can easily be recognised by pronounced horizontal cracks in the render surface caused by expansion of the mortar joints and is most common on the more exposed parts of a building, such as parapets

and chimney stacks. If brickwork is to be rendered, it is wise at the design stage to specify sulfate-resisting cement, bricks with a low salt content and a render that permits evaporation from the brickwork.

In industrial areas, there is also the risk of attack by sulfur oxides, which can attack the cementitious materials and cause discolouration of the surface.

Failure to set

If the render does not set properly, the cause can be inadequate protection during drying. In hot weather, the water can evaporate before hydration takes place. Similar problems can be caused by contaminated or old materials, or very dry porous walls, which suck the water out of the render.

Lime bloom

Lime bloom can affect most renders and is a white film of calcium carbonate which is caused by a chemical reaction during the drying process. If the render is a fairly pale colour, it should not be noticeable. On strong colours (right), it is harder to disguise.

Grinning

If the lines of the brickwork are visible through the render (particularly noticeable as the wall dries after rainfall), it is likely that the render is of insufficient thickness. It is not unknown for unscrupulous builders to build a fair face block wall and then apply a Tyrolean (wet dash) finish direct. This is rarely successful.

Cracks at the jambs

Cracks at the corner of jambs and heads round window and door openings are sometimes caused by bedding angle beads in gypsum plaster. This has a fast set and therefore seems ideal for temporarily supporting angle bead. However, in wet conditions sulfate attack can occur when calcium sulfate and cement are in close proximity. Cracks further above door openings are more likely to be caused by lintel or arch failure.

Popping or blowing

This occurs in cement renders, which include lime, usually in bagged form (hydrated lime). During manufacture, the lime is not completely slaked. In a typical case circular cracks (about 40mm diameter) appear in the render six months or so after completion.

Eventually, a cone of render is forced off the wall as small nodules of quicklime slake. Repairs are easy but difficult to disguise. Do not confuse this with wall tie repairs (right).

Structural movement

Most of the causes of cracking have been covered in the previous few pages. However, structural movement will also cause cracking. The following chapter on Cracking explains the patterns and size of cracks in more detail. This example (right) shows cracking in an end-of-terraced house situated at the bottom of a steep slope.

CHECKLIST

Element	Comments/problems
Damp problems	Look at the render and assess whether there are any details that may encourage damp penetration. These include sills and throated copings with render filling the drips, window heads without render bell mouths (to form a drip), and render running down to ground level.
Cracking	If the render has cracked try to establish whether the cracking is caused by failure of the render or failure of the wall behind it. Vertical cracks may be caused by structural or thermal movement. Horizontal cracks are more likely to be caused by sulfate attack or wall tie failure. The chapter on Cracking will provide more information on this. Assuming that cracking is only in the render, inspect the surface for signs of map cracking (shrinkage), hollow spots and bulging. Cracking can also occur along angle and stop beads. This can be caused by incorrect bead fixing and rusting of the beads. A horizontal crack just above ground level suggests that the render bridges the DPC.
	Note that any type of cracking may result in damp patches inside the house.

Element	Comments/problems
Loose and hollow renders	If substantial parts of the render have parted company with the wall, a more detailed analytical investigation is required. This will include looking at the thickness of the render, its composition and number of coats, the background material and preparation, and the presence of bonding agents or harmful chemicals.
Surface finish	Inspect the surface of the render for any signs of grinning, surface crazing, popping, lime bloom, and algae or moss growth. In pebble dash bare patches may be unsightly (they are common at scaffold-lifts) but they are usually insignificant in terms of durability. Note that some painted renders may be counter-productive in terms of damp proofing as they slow down evaporation. Clear silicon-based paints can cause similar problems. One consequence (in houses with solid walls) may be the appearance of efflorescence on the inner face of the wall (ie on the plaster).
Insulated renders	Some insulated renders comprise a thin resin-based render laid on 50mm (or so) insulation boards. These finishes are very vulnerable to impact damage, eg bicycle handlebars etc.

CRACKING

INTRODUCTION

The previous chapters have briefly described a number of potential problems that can lead to cracking or building movement. These problems are not just settlement, subsidence and heave; they also include chemical attack, roof spread, lack of restraint, wall tie failure, thermal movement, moisture expansion and vibration.

These defects have, in the limited space available to us, been dealt with in the appropriate chapters. This chapter is more of a summary. It highlights the main causes and types of cracking and explains their distinguishing features. So for example, just how do you distinguish cracking caused by wall tie failure and cracking caused by sulfate attack? Similarly, how do you tell the difference between differential settlement and ground heave?

Thousands of houses in the UK are cracked. Some cracks may affect the structural stability of a building or the safety of its occupants; some may lead to problems of damp penetration; others are insignificant. This chapter provides guidance on recognising, monitoring, and diagnosing cracking. It cannot, however, be regarded as a substitute for good professional advice. If in doubt contact a qualified surveyor or engineer.

When is cracking significant?

Most buildings have cracks of one sort or another, in the main caused by moisture and thermal movement. They are common in new houses, for example, where shrinkage in floor joists and stud partitions causes the almost inevitable cracks found around the edges of plasterboard. Cracks such as these (usually 1–4mm) are insignificant and can be filled in the course of decoration. Cracking is more significant if it is wider than 4–5mm and if it is progressive.

The Building Research Establishment (BRE) has published several guides and papers on cracking. Its *Digest* 251, produced many years ago, is still relevant. In this document, the BRE lists six categories of cracking. A simplified version of this table is produced below.

Classification of visible damage to walls
(Adapted from Table 1 in BRE's *Digest 251*.)

Category	Description of damage	Approx. width (mm)
0	Hairline cracks – classed as negligible	0.1
1	Fine cracks – easily treated during decorations. Rarely visible in external brickwork. Classed as very slight.	Up to 1mm
2	Cracks easily filled. Some external repointing may be necessary, doors and windows may require easing. Classed as slight.	Up to 5mm
3	Cracks require cutting out and patching. Repointing required. Service pipes (usually clay drainage pipes) may fracture. Doors and windows will stick. Weather-tightness often impaired. Classed as moderate.	5 to 15mm
4	Extensive repair work required, including brickwork repairs over doors and windows. Walls leaning or bulging, floors sloping. Service pipes damaged. Classed as severe.	15 to 25mm
5	Major repairs or rebuilding required. Walls leaning, beams lose bearing, windows cracked through distortion. Building unstable, shoring may be required. Classed as very severe.	> than 25mm (or several smaller cracks)

Monitoring

Where moderate, severe, or very severe cracking is suspected expert advice is required. An engineer will, among other things, assess whether movement is progressive or complete. This is usually achieved by using tell-tales (see next page). The engineer, apart from examining the crack itself, will also consider the nature and age of the building, whether there are similar problems in surrounding buildings, the nature and slope of the ground, the size of nearby trees, and the possible impact of local engineering or building works.

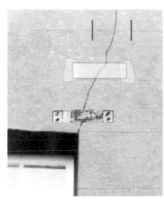

Two vertical pencil lines, say 100mm apart, can help monitor whether cracks are progressive.
Glass tell-tales bedded in mortar. If these crack, movement is still occurring – they cannot reveal much else.

Plastic tell-tales, two small sheets of overlapping plastic with gridlines on their face, can also be fixed either side of the crack.

Engineers may also use more sophisticated equipment to measure and monitor movement. These include vernier gauges (right) and strain gauges.

With regard to the cracking itself an engineer will record details regarding its:

- Position, start and finish points.
- Length, width and depth.
- Angle and shape (tapered or parallel).
- Whether the crack runs through the material or through mortar joints.
- Alignment of surfaces either side of the crack.

Foundation-related?

It is often difficult to assess whether movement is foundation-related, particularly if one is unaware of a building's history. The following criteria probably point to foundation/ground problems:

- Cracks that are visible on both sides of the wall.

- Cracks that extend below the DPC (in other words into the substructure masonry).
- Cracks that taper, either wider at the top or wider at the bottom.
- Distortion in window and door frames.
- Floors out of level.

In addition, it is important to consider other factors that may have an effect on the diagnosis:

- Long periods of dry, hot weather (in the long hot summer of 1976 claims for foundation-related defects were six times greater than 1975).
- Garden/patio works (may affect subsoil drainage or damage drains).
- Older houses, built with lime mortar, may undergo substantial movement without developing cracks.
- Many buildings that have cracked at some point in the past may now be stable (and will usually have been repointed).

The table summarises the main causes of cracking in walls. Further information can be found in the drawings on pages 152-157.

A summary of the main causes of cracking

Main causes	Comments/problems
Ground problems	• Differential settlement.
	• Mining.
	• Tree roots in clay soils.
	• Seasonal movement in clay soils.
	• Building on filled ground.
	• Shallow foundations and frost heave.
	• Broken drains, felled trees (clay soils).
	• Landslides, coastal erosion, earthquakes.
Loading	• Building alterations affecting load patterns.
	• Point loads from purlins, heavy trusses etc.
	• Thrust from poorly tied roofs.

Main causes	Comments/problems
Lack of restraint	• Walls not tied to roofs and upper floors. • Flank walls, chimney breast, party walls, internal loadbearing and non-loadbearing walls all help to restrain external walls.
Thermal movement	This is usually caused by inadequate provision of movement joints.
Freezing	In very cold conditions, frost attack may affect the floor slabs of outhouses, garages etc - most likely in sandy or chalk soils. Frost attack can also damage bricks and stone walls, however, the damage is usually in the form of spalling not cracking.
Moisture movement	• Clay bricks will expand slightly after manufacture. • Concrete blocks and sand lime bricks will shrink. • Timber will shrink in a dry, heated environment. • Some hardcores can expand if they become wet. This can crack a groundbearing floor slab. • Initial drying shrinkage of a new building. This is normally cosmetic only.
Chemical attack	• Sulfate attack – cement-based mortar joints expand – exposed parts of a building most affected. Expanding ground floor slabs may push walls outwards. • Carbonation – porous or cracked reinforced concrete most at risk. In traditional houses lintels are the element most likely to exhibit problems of carbonation. • Rusting – wall tie failure, lintels.
Vibration etc Impact damage	Traffic, machinery, quarrying, nearby piling. This is most likely to be caused by falling trees or speeding vehicles.

Cracks – foundation problems

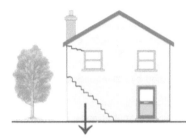

Ground shrinkage as tree extracts moisture from clay subsoil. Cracks in the masonry are usually wider at the top. Cracks usually occur in mortar joints.

Ground heave following tree removal (clay soils). Force of heave can crack bricks (not just mortar joints). Cracks in masonry are wider at the bottom. Broken drains can cause similar problems.

The crack patterns in these two examples may look similar but it is the relative crack width (top and bottom) that helps an accurate diagnosis.

Sagging – cracks wider at the bottom

Hogging – cracks wider at the top. Cracking may continue through the roof.

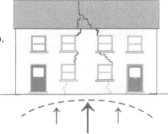

Cracks – more foundation problems

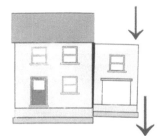

Original house foundations are deeper than new extension. Outer wall of extension is likely to drop causing cracking and rotation. If the existing roof was extended over the extension, cracking would also be evident in tiling.

The house shown in the right-hand photo was built in the late 19th century. At some time later the house was extended. There are signs of settlement and rotation; they have affected the roof, not just the walls. The gap between the walls has been filled with mortar.

House extended by adding extra storey. Settlement of the main houses can cause distortion of existing additions if they are bonded together.

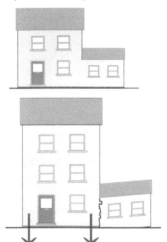

The house shown below was built on filled ground. The house developed a series of random cracks within a few months of completion. The house was eventually underpinned using concrete piles. The piles transferred the loads down to a deeper, firm stratum.

Cracks – wall problems – loading

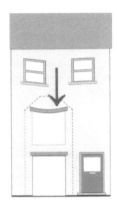

The brickwork between the windows drops as the lintel (bressumer) deflects (often caused by rot). Whether or not the wall cracks depends on the amount of movement and the nature of the mortar.

In the photo on the right you can see a massive crack where the stonework in between the windows has dropped.

Feet of rafters poorly restrained due to rusty nails, inadequate ties etc. Spreading rafters can push the wall out of plumb. Vertical cracking may occur at the top of flank walls; horizontal cracking appears on the front wall a few courses below the wall plate.

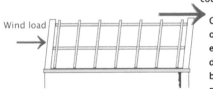

Gable end has been pushed out of alignment. Many early trussed rafter roofs did not contain diagonal bracing. This would have prevented racking.

The same problem can occur if wind blows between two gables, causing negative pressure and pulling the gables out of alignment. Nowadays, this is avoided by restraint straps.

154

Cracks – wall problems – horizontal movement

In terraces of houses, vertical cracks are often found between the downstairs and upstairs windows. This is caused by thermal movement. In long hot weather, the terrace expands slightly. As it cools, the bricks cannot pull it back into shape and it cracks. If the terrace is built in lime mortar, cracking is less likely to occur although the joints might open up.

Expansion in summer, contraction in winter

New calcium silicate (sand/lime) bricks shrink after manufacture. Here they have been used too soon – hence the filler.

The three photos below all show cracking caused by horizontal movement/forces.

Parapets are particularly prone to thermal movement. They are exposed to the elements on both sides and are often built on slip planes (DPCs). In extreme circumstances, the expanding wall can cause cracking at the quoins.

155

Cracks – wall problems – lintels

Arch failure is often a consequence of building movement. In many cases a timber lintel on the inside face of the brickwork will prevent total collapse. Many rendered buildings hide simple arches. The pattern of failure on the left suggests this is not a superficial render problem.

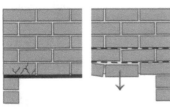

(Right) Rusting steel mesh reinforcement in bed joints allows bricks to drop.
(Left) Wrought–iron or steel lintel rusts and the expansion causes spalling and cracking of the brickwork.

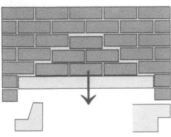

Concrete boot lintels were popular in the 1930s and 1950s. Some of them were only supported by the internal leaf. The weight of the external leaf could cause rotation in the lintel, leading to cracking in the brickwork.

Lintel section and plan

Some concrete lintels crack because the embedded steel reinforcement rusts. This is a particular problem in the concrete houses built after the Second World War.

Cracks – wall problems – saturation

In the UK saturation probably accounts for most building defects. Damp penetration aside, saturation can lead to problems of wall tie failure and sulfate attack, both of which can manifest themselves as cracking.

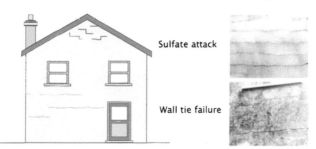

Sulfate attack

Wall tie failure

Wall tie failure will only occur in cavity walls. The expanding ties can force the mortar joints apart. The cracks usually appear every six courses. Sulfate attack is most likely to occur in the more exposed (wet) parts of a building - chimneys, top of gable walls, garden retaining walls etc. It can occur in every course.

Cracks – wall problems – renders

If a building is rendered cracks may occur that have nothing to do with wall or foundation movement. A horizontal crack just above the ground sometimes occurs if the DPC has been rendered over. Shrinkage cracking occurs where a dense render (cement rich) has been applied to a weak background. Minor vertical cracks sometimes occur where extensions join the main structure or around windows if angle beads rust. See previous chapter on Rendering.

These cracks are <u>not</u> superficial.

These cracks are superficial - in the render only.

GARDEN/BOUNDARY WALLS

INTRODUCTION

This chapter examines the construction and failure of garden walls. It does not include retaining walls. Garden walls should be inspected regularly because they are prone to failure; collapsing is not uncommon. In addition, garden walls are likely to be next to pavements, so they are a common cause of serious injury. A badly maintained wall may result in a huge personal liability: it may not be covered by building insurance if maintenance has been neglected.

GOOD CONSTRUCTION

There are some generally recognised principles of good construction. These are listed below; further details are shown in the drawings.

- Walls up to 1.8m high can be constructed using good practice as a guide. Above this height, walls should be designed by an engineer.
- The acceptable height of a wall, assuming it is correctly built, depends on its thickness. Guidance is provided by the Office of the Deputy Prime Minster (ODPM). The ODPM also produce the Building Regulations and other construction-related legislation). Wall thickness can be reduced if an adequate number of piers are provided.
- Piers should normally be provided at the ends of long runs and at both sides of openings.
- A wall needs a continuous foundation of adequate depth and size.
- Vertical movement joints should be provided where a wall meets a building and at regular intervals (typical intervals are 6m for blocks, 7.50m or so for calcium silicate bricks, 10–13m for clay bricks). Recommendations for the spacing of movement joints vary, however, from manufacturer to manufacturer. Ibstock bricks, for example, recommend joints every 6m (clay bricks). Older walls built in lime mortar (mostly pre 1950s) will not usually require movement joints.

- DPCs should be present at the base of the wall and under the copings. These should be preferably rigid materials, such as slate, tiling and engineering bricks.
- Bricks should be of good quality, preferably frost-resistant. The brick quality will also depend on exposure.
- Mortar should be typically 1:1:5–6 cement lime sand (or its equivalent if plasticiser is used) for the main part of the wall. It should be stronger, say 1:0.5:4-4.5, for the top and bottom of the wall. It is recommended that the stronger mix is used for all of the wall if the copings do not overhang.
- Copings should be frost-resistant and should, preferably, throw water away from the wall.

GARDEN WALL FAILURE

General defects

Defects in garden walls are usually fairly easy to recognise and do not require detailed discussion. We have, however, included, where appropriate, explanatory information in the checklist. We have also included a page of photographs before the checklist showing a number of common defects.

Brief guide to cracking

The drawings at the end of this chapter are adapted from the *Good Building Guide Number* 13, 'Surveying Brick or Block Freestanding Walls' published by the BRE. The guidance on cracking should be read with caution – if in doubt seek expert advice.

Garden walls will crack for a number of reasons. These include:

- frost attack in the mortar or heave caused by frost in the ground
- sulfate attack
- thermal and moisture movement
- shrinkage or expansion of new materials
- ground movement
- inadequate foundations
- tree damage - through soil shrinkage or damage from roots
- walls of inadequate width (in relation to height)
- impact damage
- alterations – such as new openings
- incorrect construction, piers not bonded in, incorrect mortar mixes, incorrect type of DPC etc.

Safe wall heights (garden walls)

Britain is divided into four zones. The further north, the lower the maximum safe height of the wall. Within each zone there may be sheltered areas – in these situations the wall height may be slightly higher than the figures given below.

Zone	1 Max height	2 Max height	3 Max height	4 Max height
Half brick	525	450	400	375
1 brick	1450	1300	1175	1075
1.5 brick	2400	2175	2000	1825
100mm block	450	400	350	325
200mm block	1050	925	850	775

If suitably sized piers are built on both sides of the wall its thickness can be reduced.

In zone 2, for example, a 1 brick wall (215mm) can be 1300mm high. If piers are attached as shown below, it can be built as a half-brick wall.

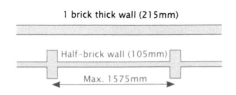

1 brick thick wall (215mm)

Half-brick wall (105mm)

Max. 1575mm

Garden walls – details

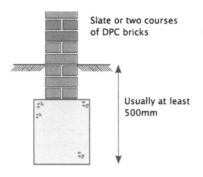

Slate or two courses of DPC bricks

Usually at least 500mm

Brick coping on tile DPC, or brick coping on high bond DPC (two brick courses required), or proprietary coping (with drips) on high bond DPC.

Typical mortar mix for main wall – 1:1:6; for copings and work in foundations, or for main wall if not protected by overhanging coping – 1:0.5:4.5.

Tooled joints are preferable. Avoid recessed or flush joints if possible.

Brick on edge coping (frost-resistant) on tile DPC.

Bricks between DPCs can be frost-resistant or moderately frost-resistant depending on exposure.

All bricks below DPC to be frost-resistant.

Thinner walls than those recommended (see previous page) are permitted if piers are added at regular intervals.

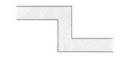

Walls can be strengthened by providing returns etc. These should be designed by engineers.

Garden walls – cracking

Cracks that are probably insignificant

Single cracks up to 5mm wide in mortar
or bricks but not in piers.

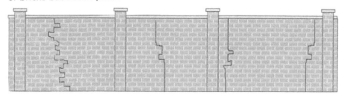

Two cracks up to 5mm but at
least wall-height apart.

Cracks to monitor or investigate

Single cracks up to 5mm   Two cracks up to   Single cracks over 5mm
within 400mm of a pier   5mm – less than   but not extending over
end.   wall-height apart.   600mm horizontally.

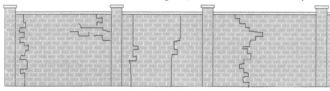

Multiple hairline
cracks.

Cracks that may require rebuilding

Single cracks over 5mm within   As above but over 600mm
400mm of a pier end.   horizontally.

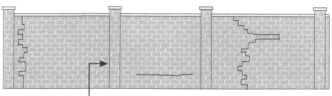

Cracks where pier   Single cracks more than
joins wall.   600mm long running right
through wall.

Based on GBG 13, BRE 1992

Garden walls – typical defects

Garden walls will often crack if they are rigidly attached to main house walls – especially if they have shallower foundations.

Frost attack (above left) and sulfate attack (right).

The wall on the right has been incorrectly repointed with a strong brittle cement mortar.

Both the walls below have inadequate foundations.

This 1 brick wall is being pushed over by tree roots growing at the base of the wall.

This half-brick wall has been strengthened with piers, but there are no DPCs (top or bottom) and the brick on edge coping is not frost-resistant.

CHECKLIST

Areas to inspect	Comments/problems
Is the wall upright?	If a wall leans it may be at risk of collapse. A half-brick wall should not lean more than 30mm, a 1 brick wall not more than 70mm.
Is the wall too high? Is it stable?	Check the height of the wall in relation to the table and map on a previous page. Remember that in sheltered areas the height can be increased slightly.
Are trees or vegetation affecting the wall?	Check whether there is any damage from climbing plants or any potential damage from nearby trees.
Signs of cracking or movement	See the guidance on previous pages. The significance of cracking depends on its width, its frequency, and its proximity to piers or other cracks. Not all cracking is significant. Expert advice may be necessary.
Ground level either side	If ground levels either side of the wall are different there may be lateral pressure on a wall which can push it over. Note this chapter does not include guidance on retaining walls.
Movement joints	If movement joints are necessary (see main text) are they sufficient and are the joints free from mortar or other debris?
Changes to openings	The addition of gates etc may have weakened the wall, particularly if extra piers have not been added.
Damage to the bricks themselves	The most common problems are frost attack and sulfate attack. These problems are probably only serious if they affect large expanses of the wall.
Mortar condition	Walls built in old lime or old and weak cement mortars may require repointing. Strong cement mortars may have shrunk away from the wall or cracked. If mortar can be scraped out to a depth of 10mm or so, repointing is necessary.

Areas to inspect	Comments/problems
Coping or top of the wall	Is the coping on a DPC, does it shed water? Most important of all, is it well bonded?
Signs of damage from rising dampness	This will usually manifest itself as staining (efflorescence) just above ground level. In very cold locations it may be accompanied by frost attack.
Correct construction	Is the wall properly bonded, both across the face of the wall and through it? It may appear to be a 1 brick wall but it could be 2 half-brick leaves laid side by side. If so are there any wall ties? Are piers correctly bonded to the wall itself? Are DPCs of a rigid type and properly bonded? A rendered wall can hide many problems, the render may also be cracked (see chapter on Rendering).
Car damage	Cars can cause minor abrasive damage. They can also push walls over, or substantially weaken the wall just by nudging it.

PLASTERING

Lime plaster

Until the 1950s most plasterwork was based on lime. Lime is made from chalk or limestone (calcium carbonate), which was (and still is) heated in kilns to drive off the carbon dioxide. The resultant material, quicklime, was then added to water (slaking) to produce calcium hydroxide. When the slaking was complete, the mix was run through sieves to screen out any unslaked lime. It was then stored in large pits where it was usually kept at least two or three months before being used to ensure the slaking was complete. The lime preferred for plastering was known as pure lime or 'fat' lime. It contained no impurities (which might speed up the set). The best limes are from Crich, Buxton and Dorking. The lime was mixed with sands, ash and other fine aggregates and then applied to the background in two or three coats. The plaster hardened by drying and by absorbing carbon dioxide from the atmosphere. This was a very slow process – it could take several months before the plaster was hard enough to paper or decorate.

The plaster was normally applied in three coats known as render, float and set. The render coat roughly levelled out the wall; the floating coat evened out the suction caused by the differing thickness of the render coat and provided a level surface for the setting coat, which gave a smooth true finish. Lime plaster shrinks as it dries (as the water evaporates the lime tends to consolidate) and to help prevent cracking of the first two coats they were reinforced with animal hair.

Typical mixes would be three parts sand (or whatever aggregate was locally available) to one part lime for the render and floating coats, and equal parts of lime and fine sand for the finish or setting coat. The total thickness would depend on the nature of the wall. On good brickwork, a total thickness of about 19mm is typical (8mm, 8mm and 3mm).

Lime could also be applied to strips of timber known as lath – this was the norm for ceiling work until the mid 20th century. On ornate ceilings the lime could be 'run' to form cornices or it could be applied in the form of pre-cast mouldings.

Some limes contain a small proportion of clay. This gives the material (hydraulic lime) a much stronger and faster set. Although this was suitable for some bricklaying mortars and renders it was not

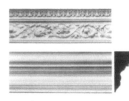

Elaborate ceiling decorations were cast in Plaster of Paris (fibrous plaster). Hundreds of styles were available. Simpler cornices and mouldings could be 'run' in wet plaster.

good for fine internal work. Fine aggregates such as fuel ash, volcanic ash and even brick dust could be added to a 'fat' or pure lime to induce a hydraulic reaction. Strong hydraulic limes are not that dissimilar to cements.

Gypsum plaster

Gypsum in its naturally occurring state is calcium sulfate combined with water. It is quarried in many parts of the UK. The material is ground and crushed to a fine powder and then heated in a kiln to produce a material known as Plaster of Paris. This is used to form ceiling mouldings but it is no use for plastering as it sets too quickly. However, by adding various chemicals the set can be slowed down – these slow-setting plasters are ideal for use on walls. On most walls two coats are usually acceptable. These plasters steadily increased in popularity during the middle of the 20th century. Cheaper manufacturing techniques (and much faster drying times) meant that they could compete against traditional lime plasters. Originally these plasters were mixed with sand on site (usually three parts sand to one part plaster for undercoats, one to one for finishing coats). Since the 1960s, they have been available as pre-mixed, pre-bagged plasters – only water has to be added on site. Nowadays, plasters contain lightweight aggregates, rather than sand. These improve the working qualities of the plaster and also improve thermal insulation. They are available in a wide range of grades to suit differing backgrounds. They are not, however, suitable in damp conditions.

Cement plasters

Gypsum plasters are unsuitable where there are damp walls or where walls have to be replastered following DPC installation. They readily absorb water and will eventually break away from the background. Furthermore, gypsum plaster cannot stop the passage of salts caused by migration as water evaporates from the wall. Some of these salts occur naturally in building materials, others are drawn from the wet ground. These salts, depending on their chemical make-up, can either appear as efflorescence on the surface of the wall (a white fluffy powder) or as well-defined damp patches. Following DPC installation it is usual to use render and floating coats of sand and cement containing a waterproofer or salt retarder to keep the salts away from the face of the plaster.

It is quite acceptable to use sand cement plasters on new blockwork although great care is required to minimise the risk of shrinkage. Lime or plasticisers should normally be added and the strength of the mix should take into account the key and suction of the background.

PLASTERBOARD

Plasterboard was introduced some 80 years ago and first became popular as a ceiling finish, gradually replacing timber lath ceilings (shown on page 171). For the past 30 years or so, it has also been used as a dry lining for walls.

Plasterboards comprise a layer of gypsum covered both sides, and along the long edges, with stiff paper. They are available in range of board sizes and thicknesses. The boards are usually self-finished (see graphics) although many of them can be skimmed with a thin coat of gypsum-based board plaster. Some grades include a bonded thickness of insulation and/or a vapour check layer. The boards are fixed using dabs of adhesive, timber battens or metal channels.

Until the 1990s most boards were 9mm or 12.5mm thick. These were usually fixed with galvanised plasterboard nails. The timber studding or joists would require extensive noggings to prevent the boards from flexing. Nowadays 15mm boarding has become more popular. When fixed with screws additional noggings are not normally required.

Nowadays, boards are usually fixed to dabs or metal battens. The former is the simpler of the two systems. It comprises a series of adhesive dabs, typically 50–75mm wide and 250mm long, applied by trowel to the wall. Three 'columns' of dabs are normally required per board, with horizontal dabs between the columns at ceiling level, and a continuous band of adhesive at skirting level. When the dabs are in position, the board (cut 15mm short of wall to ceiling height) can be pressed and tapped into position, tight against the ceiling.

The basic procedure for jointing is as follows: apply jointing compound along tapered joint, bed joint tape into compound, apply jointing compound along top of tape and feather out. Repeat this once or twice. Internal and external angles can be finished in the same way (external angles usually require a special angle tape). The taping and jointing exercise can be done by hand or by hand-held machine.

When the boards have been taped and finished it is good practice to give them a coat of primer/sealer. These products vary from manufacturer to manufacturer. Their purpose is to even out the difference in texture and absorption between the board surface and the jointing compound. Without it the joints may 'grin' through the paint. Some primer/sealers also help subsequent removal of wallpaper.

Ceilings

Timber lath ceilings

The laths were strips of fir (sawn or split) nailed to the soffit of the joists. The lath was usually finished with three coats of lime plaster. It was very time-consuming in terms of both labour and drying.

Joists shown here at 450mm centres

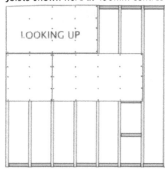

LOOKING UP

1800 x 900mm plasterboard 12.5mm thick. Some boards are 2400 x 1200mm.

Boards should be laid with staggered joints. Perimeter noggings are required for 12.5mm boards but not centre noggings. If joists are are at 600mm centres (ie normal for trussed rafters) centre noggings are also required. If 15mm plasterboard is used, noggings are not required at all (450 and 600mm centres).

Plaster lath (various trade names) – boards are usually 1200 x 600mm. Plaster lath is laid with staggered joints and with a 3mm gap between boards. Centre noggings and taping are not required. It is designed for a plaster finish.

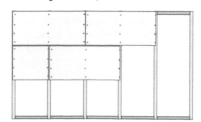

Section through lath joint

A coat of sealer should be applied to the boards before they are painted or papered. This evens out the suction between the boards and the joints and seals the paper surface (providing a better paint finish).

Sealer is also required before applying wallpaper so that the covering can be removed for redecoration without damaging the wallboard surface.

The finished boards accept most types of paint and textured coatings. Matt finishes are the best; gloss or sheen finishes tend to highlight any minor surface irregularities.

External angles can be covered with paper tape, or very fine angle beads bedded in jointing compound and feathered-out as necessary.

Ceilings

For the past 50 years or so plasterboard has also been used to line ceilings. Typical details are shown on the previous page.

PARTITIONS

Introduction

Most houses have a combination of loadbearing and non-loadbearing partitions. Loadbearing partitions are, nowadays, built in blocks or timber studding. In the past, they have also been built in brickwork. They usually support upper floor joists and, in traditional roofs, part of the roof structure. In modern houses with trussed rafters loadbearing partitions are not required on the upper storey because the trussed rafters do not need any intermediate support.

Non-loadbearing partitions can be built from the materials mentioned above but are nowadays more likely to be formed using proprietary plasterboard systems (see page 174). In the first half of the 20th century, non-loadbearing partitions were usually built in timber studding covered with lath and plaster or in lightweight blockwork. A number of lightweight blocks were available made from a mixture of cement and a variety of industrial wastes; clinker and breeze were two common ones. Clinker and breeze walls were light enough to be supported by a timber floor although additional joists were often required where the wall ran parallel to the joists (ie directly under the wall). The next two pages show a variety of partition types.

Stud partition – typical construction c1900

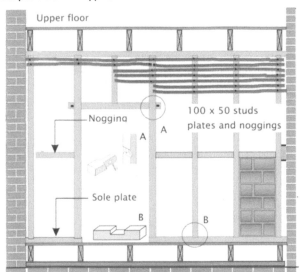

Upper floor

Nogging

100 x 50 studs
plates and noggings

Sole plate

Stud partitions were usually covered with timber lath (fir strips) and then plastered with three coats of lime plaster. A brick on edge infill gave improved sound insulation and fire protection (usually found on the ground floor only).

Stud partitions are still common today. They can be used in alteration work as well as new-build. In modern construction the studding is likely to be covered with 15mm plasterboard; self-finished or skimmed. If the partition separates rooms, it should normally include a sound deadening quilt. Additional noggings can be added to support radiators and sockets etc.

Laminated partition

This usually comprises three layers of
plasterboard bonded together. It is not
loadbearing but does provide quite good
sound insulation.

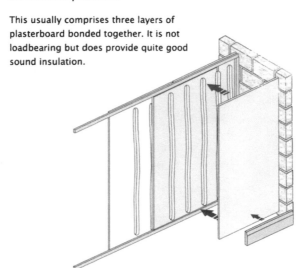

Cellular core partition

The panels comprise two layers of
plasterboard with a cardboard cellular
core. This, again, is a non-loadbearing
partition.

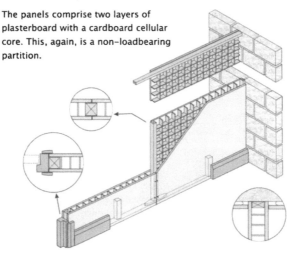

DEFECTS

Plaster

Efflorescence

Efflorescence can be a minor problem on new work. A very small percentage of efflorescent salts may exist in gypsum plaster, and larger amounts will occur in bricks, some concrete blocks and

undercoats based on cement. During the drying process, these are carried in solution to the surface and can damage decorations. They can be brushed off as they appear, and should disappear after a few weeks. If they do not subside it could indicate a problem of penetrating or rising damp.

Dampness

Most plasters will break down fairly quickly if they remain damp for long periods. Gypsum plasters are

particularly at risk and they are generally unsuitable for use on solid walls other than as finishing coats to sand/cement plasters. The picture shows two-coat gypsum work on the inner walls of a basement applied as part of a repair. It lasted only a few months.

Fast set

Gypsum plaster that is fresh from the works can sometimes still be hot. If it is not allowed to cool before use as a fast set can result. Similarly, dirty water or mixing tools can speed up setting time. If the set is too fast the plaster will not bond properly to the background.

Dry out

In exceptionally hot conditions, gypsum plaster can dry too quickly. The water, which is required to return the plaster mix to its dehydrate state, may evaporate, leaving the plaster soft and powdery. In normal weather conditions a similar problem can occur through incorrect choice of plaster, or poor preparation of background. On very porous materials with high suction, such as aerated concrete blocks, it is often necessary to seal the wall with a proprietary sealer. If this is done the background may suck most of the water out of the plaster mix before the chemical set has taken place. The plaster cannot set properly and usually has a rough powdery surface.

Grinning

If plaster is applied in less than the recommended thickness, grinning may occur. For example, on a concrete block wall laid in a weak mortar the plaster will dry more quickly over the porous joints. If the plaster is not thick enough, the lines of the joints will be visible through the plaster. Wallpaper does not always hide the problem and the only acceptable solution may be to hack off and replaster.

Shrinkage cracking

This is unlikely to occur in lightweight pre-mixed gypsum plaster as it expands slightly on drying. It can, however, occur in properties with cement and sand undercoats. These shrink slightly as they dry and if the thin gypsum finish is applied before the shrinkage is complete, cracking will occur. Treatment of these fine cracks is extremely difficult (apart from hacking off and replastering) and such surfaces should be finished with wallpaper. The same pattern of cracking (but on a much smaller scale) can occur if finishing plasters dry out too quickly.

Cracking caused by thermal and moisture movement

This is most commonly caused by the drying shrinkage of concrete blocks. The cracks are usually easy to recognise as they tend to follow the mortar joints. The cracks can be cut out and filled, but if the underlying problem is not resolved they may reappear. Problems of thermal movement are unlikely to manifest themselves in housing because the 'runs' of blockwork are usually of limited length.

Other cracking

If you come across other types of cracking during an inspection try to work out if the cracking is just in the plaster, just in the inner leaf (if it is a cavity wall) or right through the wall. You should also think about the nature of the original construction; this often offers clues to the cause. If the cracks are wider than 3mm or so and the cracking is diagonal (cracking or stepped), consider settlement or subsidence. Horizontal cracks, which run through both leaves (if it is a cavity

wall), could indicate wall tie failure. In a solid wall, horizontal cracks right through the wall might indicate sulfate attack (see the chapter on Cracking). Vertical cracks (in the plaster only) above light switches or sockets probably indicate insufficient plaster cover over trunking. Horizontal cracks just below the upper floor ceiling may suggest the joint between the blockwork and wall plate has not been covered with metal lath. It might also be caused by differential movement of the brick and block leaves (causing slight rotation at the top of the wall) or even failure to strap the wall plate down.

Plasterboard (drylining)

This can be affected by the following:
- dampness
- impact damage
- incorrect nailing (nail popping)
- inadequate background support (not enough dabs, stud centres too great)
- incorrect taping
- boards too thin.

Ceilings

Lath and plaster ceilings can easily last a 100 years. However, they can fail prematurely for a number of reasons. These include:

- Laths nailed too closely or too far apart.
- Rusting of nails or rotten laths.
- Vibration causing the plaster to lose its bond with the lath – this can be caused by undersized joists, lack of strutting etc. This usually manifests itself as 'map' cracking.

In the post-war years a number of ceilings were lined with fibreboard or various asbestos cement boards. By the late 1950s most ceilings were finished with plasterboard. It is unusual to find defects in the plasterboard itself, assuming it has been stored in dry conditions, and most problems can be attributed to incorrect fixings, or badly prepared background support. Boards may bow or sag if the fixing centres are too far apart, if the boards are allowed to get damp, or if noggings are omitted. Loose boards are often a result of insufficient pressure being applied to the board as it is nailed in position. Nail popping can occur if the incorrect nails are used (screws are preferred nowadays), or as a result of drying shrinkage in a timber background. Major cracks in the joints are also likely to be a result of shrinkage, or lack of strutting in a timber floor or study partition. Hairline cracking (mostly caused by drying shrinkage) at the junction of the ceiling and the walls nearly always occurs. It is usually nothing to worry about.

Partitions

There are a variety of reasons for partition failure, but the usual causes are errors in detailing or poor workmanship rather than material failure. As they are not exposed to the weather, problems of damp penetration should not occur, although on ground floors they may be subject to rising damp where they have not been correctly isolated from the substructure. The list below is a general guide to the most common areas of failure.

Brick and block partitions

Settlement of a loadbearing partition

This is usually caused by inadequate support or ground movement. It can be recognised by cracks in the wall and torn wallpaper at junctions with external walls and non-loadbearing partitions. In addition, doors may be difficult to shut as the linings or frames may distort. Settlement often causes tearing of the DPM and DPC, in which case the bottom of the partition may be damp. Repairs can be expensive as some form of underpinning may be required.

Lateral movement

This is usually caused by inadequate tying-in to external walls or other partitions. Methods of doing this have been mentioned earlier. Other problems include uneven thrust from strutted roofs, which can result in vertical as well as lateral movement. Finally, heavy shelving can overstress the thinner block and brick partitions, especially if the shelving is adjacent to door openings, ie the weakest part of the wall.

Cracking

The problems above will obviously cause cracking and, in fact, the nature of the cracking can give a clue to its cause. However, do not assume that all cracking is of structural significance. Incorrect mortar mix preventing even thermal movement of the wall, shrinkage of wet blocks and incorrect mix of plaster all can cause unsightly cracks without excessively affecting the strength of the wall.

Stud and plasterboard partitions

Cracking at joint with external wall

This can be caused by general shrinkage or inadequate fixing. Timber stud partitions are particularly prone to shrinkage, especially if the timber has spent some time unprotected on a wet site. The battens or studs can be fixed to the internal skin of blockwork by the use of nails, or plugging and screwing (some aerated blocks will take screws directly). The use of screws is not as fast as nailing but does provide a more secure fixing.

Cracking at joint with ceiling

This is usually caused by the slight shrinkage of the floor joists above the partition, or loading deflection causing sagging of the floor on which the partition stands.

Twisting of partition

This is usually caused by the excessive weight of shelving, heavy doors or insufficient noggings between individual struts.

Buckling of partition

This can be caused by overloading a non-loadbearing partition with shelving. Where non-loadbearing partitions are used on upper floors they should not be erected until the trussed rafters have been tiled. The trusses deform slightly under load.

Cracks around door

These often occur where fairly light partitions are supporting heavy doors. Self-closing fire doors, for example, are quite heavy. Over a long period, continual slamming of the door can crack the plasterboard joints or even loosen the plasterboard itself.

CHECKLIST—WALL PLASTER

Element	Comments/problems
Plaster type	Assuming the wall is covered with insitu plaster (ie not boarding) try and work out whether the plaster is cement-based, lime-based or gypsum-based. New-build work from the late 1960s onwards is most likely to be finished in two coats of lightweight gypsum plaster. Anything before the 1950s is likely to be lime-based.
Cracks	Look for any signs of cracking. Refer to the previous section on defects and the chapter on Cracking to assess whether the problem is cosmetic or more serious.
Loose plaster	Look (and listen) for any patches of hollow plaster. If plaster sounds hollow but is not loose and there are no signs of cracking it can probably be ignored. Do not forget that where plaster has been applied to some partitions and most drylining systems it will always sound hollow when tapped.
Staining	Staining in the form of efflorescence usually indicates the presence of dampness. In new work this could be construction water drying but in older properties it is likely to be penetrating or rising dampness. If the plaster is covered in black-spot mould suspect condensation (and look at the chapter on Damp).
Powdery surface	If gypsum plasters have a rough powdery surface it suggests that they were not mixed or applied properly or that the background suction was too high.

CHECKLIST—DRY LINING

Element	Comments/problems
Signs of movement	Look at all dry-lined walls very carefully for any signs of movement. Dry lining is fairly durable as long as it is not mistreated. Where boards are mechanically fixed, they are unlikely to part company from their timber battens or steel

Element	Comments/problems
	channels. Boards that are secured on dabs are more prone to movement, especially if heavy shelves etc have been fixed to the boards (rather than to the wall behind).
Surface finish	There are three or four common reasons for a rough or uneven board finish (assuming the boards are self-finished). These are: • The quality of the original taping and jointing. • Boards that have not be sealed prior to painting. • Damage caused by careless removal of wallpaper (exacerbated if the boards were not sealed before papering). • Badly repaired impact damage. If the boards have been skimmed, these problems should not occur.

CHECKLIST—CEILINGS

Element	Comments/problems
Lath and plaster ceilings	In houses built before the Second World War the ceilings are likely to be lath and plaster. Have a good look at the ceilings to see if there are any signs of loose plaster or bulging. If you are unsure have a gentle prod with a broom to see if there is any movement. If you suspect a problem, it may be necessary to investigate further. If you can remove an upstairs floorboard you should be able to see the lath and assess the quality of the plaster bonding. From within the roof space you should be able to remove the insulation and do the same.
Plasterboard ceilings	You need to assess the following. • Are there any hairline cracks in the centre of the ceiling? If so, do they suggest that the boards have been laid without staggered joints. • Signs of bowing – either in the middle or around the edges. • Evidence of nail popping. • Cracking or ridging which could be caused by lack of strutting.

CHECKLIST—PARTITIONS

Element	Comments/problems
Partition type	The first stage is to identify the partition type and assess whether or not it is loadbearing. If the house has a traditional 'cut' roof at least one of the walls upstairs will be loadbearing, carrying, for example, ceiling joists and/or purlin struts. Trussed rafters span from external wall to external wall without the need for intermediate support. In most houses, some of the partitions downstairs support the upper-floor joists.
Bulging and bowing	Check whether the partitions are plumb. Signs of bulging and bowing indicate excessive loading, either vertically or horizontally. The former could include the additional loading caused by recovering a roof. The latter is most likely to be caused by heavy doors or heavy shelving.
Support for partition	If doors are difficult to shut or if the floor below the partition bows slightly try to establish whether the partition has adequate support. Where joists run parallel to partition, for example, the joist should be directly below it and of adequate size. Where roofs have been recovered or where extensive internal alterations have taken place the foundations of loadbearing partitions may be inadequate.
Cracking	This can be a result of building movement but is more likely to be caused by: • overloading • inadequate fixing to other walls • slamming heavy doors • inadequate floor structure (see Floors) • problems with finishes (see above).

WINDOWS

INTRODUCTION

This chapter describes the history and development of domestic windows. It includes common materials, window styles and glazing methods, together with maintenance failures and common defects. Windows should fulfil a variety of functions (light, ventilation, thermal and noise insulation) and they have a huge impact on a building's appearance and character.

In recent times there has been very effective marketing by replacement window installers. Frequently such replacement cannot

be justified on technical, sustainability or aesthetic grounds. The photo shows a pair of Edwardian terraced houses. The left-hand house with replacement uPVC windows and doors, the right-hand one with original softwood windows and doors.

Materials

Timber

Since the 18th century, softwood has been the most popular material for windows. Some windows are made from hardwood, but these are not very common in domestic construction. Softwood timber is:

- cost effective
- easy to work, joint and mould into complex profiles
- easy to undertake piecemeal repairs
- has good integral thermal insulation
- a potentially renewable material.

However, timber is:
- Natural and variable, it can be difficult to guarantee consistent quality.
- Prone to movement and decay in the presence of moisture.
- Needs to be frequently redecorated.
- Sometimes required to be chemically preserved to ensure reasonable durability.

Plastic

Plastic, or more correctly uPVC (unplasticised poly vinyl chloride), has been used for windows since 1945 but it is only since the late 1970s that large numbers of uPVC windows have been made and installed. They consist of extruded plastic sections that are heat welded together to form windows. Early sections were very bulky and were so weak that they required metal or timber reinforcement. Improved extrusion design and manufacturing techniques today mean that most domestic plastic windows no longer require internal reinforcement. However, the extruded sections have to be substantial (ie chunky) to provide security and to support the glazing.

uPVC:
- is not affected by moisture
- has reasonable thermal insulation qualities
- is relatively durable and requires reduced redecoration compared with timber.

However:
- there is uncertain long-term durability
- there are significant environmental concerns
- ongoing repairs can be difficult
- the surfaces of plastic windows need to be regularly cleaned.

Metal windows

Steel windows were developed during the early part of the 20th century. Until 1945, the only protection was paint; since then most windows have been galvanised. High-performance steel windows are 'thermally broken', where the high conductance of the steel is reduced by having an insulated construction break between the outer surfaces.

Steel:
- has a consistent known quality
- is unaffected by moisture movement
- is cost effective.

However, it:
- Can corrode if protection is damaged or breaks down and therefore it requires regular and frequent redecoration.
- Is a poor thermal insulator.

The picture on the left shows a galvanised steel casement window. It has two side-hung opening sashes, a single top-hung opening sash and a fixed sash. The metal window is fixed in a softwood frame.

Aluminium

Aluminium windows are not very common in housing. They have been manufactured since the 1950s and consist of extruded sections mechanically jointed or welded together. Early versions were not thermally broken and suffered from condensation. There are a variety of decorative surface finishes commonly used and they should be durable. The photo shows an aluminium, sliding sash window.

Aluminium:
- has a consistent known quality
- is unaffected by moisture movement
- has a long decorative maintenance cycle.

However, it:
- is relatively expensive
- is a poor thermal insulator if not thermally broken.
- there are concerns regarding the environmental cost of production.

Window types

Casement windows consist of a frame that supports a number of hinged opening or non-opening glazed sub-frames. These are referred to as sashes. In the UK (as opposed to the rest of Europe), the sashes typically open outwards. The casement window can consist of a single opening or non-opening sash or it can be divided up using horizontal members (transoms) or vertical members (mullions).

Typical pre-First World War casement windows were a cruciform design, with a single transom and a single mullion creating four sashes within a single casement frame. Early casement designs had the sashes set into a single rebate in the window frame. By the 1960s both the frame and the sashes were rebated to improve weather resistance. Whatever the material used casement windows remain the most common window type in modern housing. The photo above shows a new replacement softwood casement window. One opening sash and one fixed sash. Note that the window is fixed close to the external wall face to ensure the timber sill projects beyond the wall.

Sliding windows consist of a solid or box frame that carries a pair of opening glazed sashes. The most common form of sliding window is the timber, double-hung sash window, which dominated window production in the 18th, 19th and early 20th centuries. They consist of a pair of vertically sliding glazed sashes that open and close via a series of cords, pullies and counterbalancing weights. Until the late 19th century, with the industrialisation of glass production, each of the sliding sashes was divided up by glazing bars into a series of smaller panes. Early glazing bars were relatively thick (30mm or more in the 1700s, 15mm or so by the Regency period). By the late

Victorian period glazing bars were no longer a necessity; most sashes had large single glass panes. The photographs at the bottom of the previous page shows two softwood, double-hung sliding sash windows. The window on the left is a replacement in the early 18th century style with thick glazing bars, a timber sill and a stone subsill. On the right is an early 19th century sash window with thin glazing bars. Note there is no timber sill: the bottom sash closes directly onto the stone subsill. Both window frames are set back from the building face and are protected by a masonry rebate.

Durability

In timber windows, the key issue is the avoidance of moisture entering the wood, this is dependent on:
- quality of the timber
- quality of the joints
- preservative treatment
- decoration.

Timber pre-1945 tended to be of much better quality than that available today. Timber today tends to have a higher proportion of sapwood and is also less dense. Such timber will have a greater propensity to absorb moisture, be affected by associated movement and more at risk to fungal and insect attack. Pre-1950s windows had mortice and tenon joints, which provided good mechanical resistance. Nowadays, comb joints tend to be used; in these joints there is greater reliance on the quality of the glue and they offer reduced mechanical resistance. Much of the softwood now used in window production is treated with chemical preservatives to improve durability. There are increasing concerns about the use of highly toxic chemicals in construction. The more targeted use of slow-release preservatives has become increasingly used.

In timber windows the external surfaces, particularly any end grain timber and horizontal surfaces, require regular protection (generally by applying paints). Oil-based paints are the most commonly applied finish. These are relatively easy to apply, although care in selection and workmanship are essential. Such paints may or may not be microporous. Microporous paints allow moisture vapour to evaporate from the timber while stopping raindrops from penetrating the surface. The non-microporous paints tend to be rigid once set and cannot withstand minor moisture and thermal movement of the timber. Cracking of the paint finish is inevitable. The cracks allow moisture into the timber that will then expand. The impermeability of the paint restricts subsequent evaporation and exacerbates cracking of the finish. High-performance stain systems have been developed over the last 30 years or so. These tend to be microporous and, as such, provide a more durable finish than the traditional oil paint finish.

Early steel windows relied on the maintenance of a painted finish to avoid corrosion. Later versions (post-1945) were hot-dipped galvanised, which provided good protection. Durability of steel windows is dependent on the avoidance of corrosion and therefore maintenance of the decorative finish is critical.

The typical finishes to aluminium windows, such as anodising and polyester coatings, can break down over time but the material itself is relatively durable and corrosion-resistant. Some aluminium windows were self-finished. Over time, this so-called 'mill finish' can slightly corrode, though not a defect, it can be unsightly.

The maintenance problems of timber windows, are well understood and well documented. uPVC windows, by comparison, are relatively new and problems may arise in the future. Nevertheless, they appear to have reasonable durability. Some manufacturers claim up to 40 years, as long as frequent cleaning is undertaken, while others suggest a life of 25 years. uPVC is known to suffer from brittleness and discolouration caused by UV radiation and a number of paint manufacturers produce paint specifically for uPVC windows.

Weather proofing

This includes:
- The quality of the detailing ensuring that water and wind-driven water is shed off of the window.
- Weather stripping to avoid casual draughts and wind-driven rain entering the property between frame and sash.
- Detailing and location of the window within the thickness of the wall.

All the horizontal parts of windows (frame and sashes) are particularly vulnerable to saturation, water ingress and rot. The lowest horizontal member of a window frame is known as a sill. Sills should be designed and fitted to ensure that water runs off their surface and drips off away from the body of the wall. Sill surfaces should be weathered (so water runs off, away from the window).

Ideally there should be steps within the sill to avoid water being blown back up the sill, and the underside of the sill should have a throating. This breaks the surface tension of water running round the edge and along the underside of the sill. The groove should run parallel to the front edge of the sill, be continuous and, ideally, should be 40mm away from the face of the wall enabling the drips to be shed away from the body of the wall. Subsills of stone, concrete and tile are frequently used and this enables the use of windows without the need for excessively wide sills. Clearly, subsills should also be sloped and should have anti-capillary grooves.

With casement windows, the sash is fitted into a rebate in the frame. In early timber windows, this single rebate in the frame provided limited resistance to draughts and wind-driven rain. Since the 1970s, most timber casements and all metal and uPVC windows had a double rebate to increase weather resistance. Not only was there a rebate in the frame, but also the sashes themselves had a rebate that fitted around the frame and provided increased wind resistance.

Weather stripping is now commonly fitted to the vast majority of new windows. Weather stripping not only avoids casual draughts and associated heat loss but, if properly, designed, can also prevent wind-driven rain entering the property. For timber windows, which suffer from a higher rate of material movement, weather stripping is essential. For older pre-1990s timber windows, where integral weather stripping was relatively rare, there are many retro-fit weather-stripping systems available.

The fixing of windows within the thickness of the wall can have a significant impact on their weather resistance and hence their effectiveness and durability. Following the Fire of London, building legislation insisted that timber window frames should be located within a masonry rebate in order to provide fire resistance. This half-brick thick rebate also provided good weather protection to the window. Since the advent of the cavity wall, windows have been rarely provided with the same degree of protection as the window frame is rarely located within a masonry rebate. The most frequent location has been to fix the window directly to the masonry jambs and towards the external face of the wall. This avoids the need for extra wide sills that project beyond the building face or the use of a subsill. A flexible joint is required to ensure that the gap between the frame and the masonry is weatherproof.

Maintainability

Because of the impact of moisture on timber windows, they require regular and frequent inspection and redecoration to maximise longevity. Redecoration itself can cause problems in operation as the paint coats build up of over time: some easing and adjustment will be necessary. It is possible and relatively simple to carry out piecemeal repair to timber windows. Fitting replacement ironmongery is also relatively straightforward.

Steel windows require decorative protection to ensure that corrosion does not occur. Piecing-in repairs are difficult and very rare.

Aluminium windows require occasional redecoration: anodising and polyester coating have a 10–20 year redecoration requirement. Small piecemeal repairs are difficult and, again, rare.

Although uPVC windows will not require such frequent redecoration, they do need frequent cleaning. Minor repairs can also be a problem. Piecemeal repairs and minor replacements are not easy and hinge and locking mechanisms can become obsolete, making it difficult to find suitable replacements. This can result in premature replacement of whole windows.

Glazing

Until the 1980s most windows were single-glazed, where a single sheet of 4mm thick glass was fixed into the aperture in the sash. Single glazing provides little thermal insulation as glass is a good conductor of heat. The low surface temperature of the single glass sheet, often below dew point, enables condensation to occur on the inner surface of the glass. This can be advantageous as the single glazing can act as a dehumidifier, removing moisture vapour and reducing the likelihood of condensation elsewhere within the building. However, condensation can cause a breakdown in decorative surfaces, affecting both the window itself and the surrounding plaster and walls. In single-glazed timber windows condensation can cause wet rot in the lower horizontal sash and frame surfaces. In ungalvanised steel windows, condensation can lead to rust and frame distortion of frames. Saturation will also reduce the thermal insulation qualities of the surrounding building elements.

Single glazing in timber windows is usually retained in the sash using 10mm long, non-headed, wedge-shaped nails (glazing sprigs) to provide a mechanical fixing. The glass is also set on a backing bed of linseed oil putty, which is also used to form a bevelled front seal. Putty will remain flexible and waterproof if it is protected from drying out, hence the importance of painting new putty a few weeks after application once the surface has hardened. Unpainted glazing rebates in timber windows can also absorb the linseed oil from the putty and hence cause hardening, shrinkage and cracking in the putty, rendering it ineffective and allowing water to enter the

timber. The use of putty to retain glass in steel windows is also very common. Fixing glazing by retaining the glass using small section timber beads, as opposed to putty, became common in the 1970s. The photo shows condensation pooling on the bottom rail of a single-glazed window during the winter.

Multiple glazing, commonly two sheets of glass – hence the term double glazing – consists of sheets of glass that trap a layer of air between them. Still air is an effective thermal insulator. By trapping the air, heat loss through conduction is reduced. Effective multiple glazing raises the surface temperature of the inner glass sheet, which, in turn:
- avoids condensation occurring on the inner surface of the glass
- reduces down draughts created by single glazing
- reduces radiant heat loss
- reduces convection heat loss.

The width of the air space is an important factor, with the optimum gap being 15–20mm. Too narrow a gap and thermal insulation is reduced, too large a gap and convection currents can be established, which will lead to heat loss, however it is accepted that reducing the gap to 12mm will not have much impact on thermal performance. In the UK, early double-glazed units consisted of two 4mm glass sheets with a 6mm air gap (known as 4:6:4). Nowadays, a standard unit will consist of two 4mm sheets, with a gap of 12mm between sheets (4:12:4). Recent developments and more stringent regulation has seen the development of glass with special coatings of metal oxides that improve thermal efficiency of the glass itself. Other developments include the use of various gases in the gap between glass sheets, which further reduce heat loss, and triple glazing.

Keeping the air gap sealed is vital if the thermal insulation qualities of the unit is to be effective. Poorly formed seals, cracked glass, a lack of or ineffective desiccants in the spacer bars between the glass sheets, can all cause air leakage. Another common problem is the chronic saturation of the seals at the edge of the glazed units through poor glazing details. Condensation occurring on the internal glass surfaces within the unit is a common symptom of a failing seal.

Almost all new windows have double-glazed units. These are most commonly fixed with a series of profiled beads. The detailing of the beadings is critical. The thickness of the bead should be enough to cover the seals on the glazing unit to avoid their degradation through UV radiation. They must retain the glazed unit, encourage water shedding and enable the drainage of any moisture that gets behind the beads into the glazing rebates. The most vulnerable section of the glazing rebate is the lowest horizontal rebate. Here, edge spacers should ensure that the unit's seals do not rest on the bottom edge of the rebate where they could sit in any trapped moisture and the bead should be detailed to enable drainage. To allow for thermal movement, flexible spacers are required at points on the edges of the unit.

Ironmongery

Casement windows require hinges for the opening sashes. These can be in steel, brass or chrome/steel. The latter materials avoid early corrosion. Hinges take a lot of wear and should be occasionally oiled to enable ease of operation and to extend durability. Nowadays, most casement windows are fitted with friction stays, which enable the window to be left open in high winds. There will also be a need for a lever to secure the sash closed and back against the weather stripping in the rebates. In traditionally constructed timber sash windows the counterbalancing weights were secured to the sashes via cords and pulleys. Today, a popular alternative is a spiral balance system, which does away with the need for cords and pulleys. The spiral balance systems do need occasional maintenance and retensioning. There is a lack of standardised ironmongery in new window production, and this can make replacement of damaged and defective items problematic. Sometimes, the lack of ironmongery means the whole window has to be replaced.

Security. Windows are a vulnerable entry point in houses and should provide security against forced entry. The standardised use of multi-point espagnolet dead locking security has become standard in most modern windows, regardless of the material used. Windows installed in the past tend to have far more rudimentary security locks. The more sophisticated security locking devices are fitted during manufacture and are difficult to retrofit. There is a wide range of post-manufacture security devices available. Fitting these to metal and uPVC windows is more complicated than fitting them to timber windows. When fitting improved security, it should be remembered that windows may have to provide a means of escape in case of an emergency.

CHECKLIST

Areas to inspect	Comments/problems
Surface finishes	Timber windows, both soft and hardwood, require regular repainting to minimise saturation and maximise durability. Check for: • Cracking of paint finish, paying particular attention to joints in frames and sashes and to horizontal elements. • Integrity of paint/stain finish to putty/glazing beadings. Check steel windows for integrity of decorative finish. In particular check for: • Signs of rusting, symptoms include raised swollen surface and orange/red staining bleeding through paint finish. Particular attention should be paid to lower horizontal sections. • Integrity of paint finish to putty/glazing beadings. Check aluminium window surfaces for signs of unsightly corrosion and integrity of surface finish. Check for surface corrosion and/or a build-up of dirt and pollution in uPVC windows.
Is the putty intact?	Inspect for cracking or missing putty, particularly on horizontal sections where the impact of missing/failing putty is potentially greater.

Areas to inspect	Comments/problems
Are the beadings secure and waterproof?	Are the beadings secure, decorated and can water drain effectively from the horizontal beads?
Structural integrity of sill	Check the integrity of sill in timber windows with a thin bladed knife. Check carefully the integrity of the joints between horizontal and vertical members. Check the underside of the sill for soft patches.
Under sill grooves	Check that the window sill: • Is shaped and fixed so that it effectively sheds water away from the window. • Has an effective anti-capillary drip. • Projects well beyond the face of the wall (by 40mm at least). If the window opening has a subsill, check: • The subsill is sloping to shed water effectively and check for an effective underside anti-capillary drip (throating). • How the window sill sits onto the subsill. Ideally, the window should be located so that its sill is slightly raised above the surface of the subsill to enable water to be shed from the sill onto the subsill. If this is the case, check for good waterproofing detail between the window and subsill. If the window sill is directly positioned on top of the subsill, water can be retained under the window sill and cause saturation and rot in timber windows.
Operation	Check: • If the windows open and close properly. Does any decoration impede opening and closing; are the sashes and frames square? • Condition of hinges, is there rust visible? Do the hinges operate smoothly? • Friction stays in casement windows. Are they adequate and do they operate effectively? Is there any sign of corrosion? • For broken sash cords.

Areas to inspect	Comments/problems
Security	Check: • To make sure that the window can be secured closed and secured open. • The locking arrangements on all windows and particularly those on ground floors. Bear in mind that windows can be a means of escape from the house in emergencies.
Glazing	Check: • For any cracks and damage to the glazing. • For condensation in multiple-glazed units.

STAIRCASES

INTRODUCTION

Staircases should enable safe access between floors in houses. Some 200,000 accidents occur each year on domestic stairs, including almost 600 fatalities. There has been significant development and recording of good design and construction practice in the last 200 years, culminating in statutory guidance in the Building Regulations. This chapter examines some of the key design issues related to the safety of staircases, describes the design and construction of typical simple domestic timber staircases and considers the common defects relating to both design and construction.

Design issues

Many existing domestic staircases will not meet current Building Regulations (they do not have to). This is not to say that they will all be poorly designed, constructed and installed. However, examining the current requirements is helpful in identifying potential dangers.

There are three key issues that must be considered:
- safety in use
- side and landing protection
- the stairs as a means of escape.

Safety in use *(largely derived from current Building Regulations)*

- Going and rise. These are not components but dimensions. The total rise is the measurement from the lower floor to the upper floor. In order to design and make a staircase, the total rise needs to be divided into an equal number of individual rises. The total going is the total horizontal measurement available for the staircase, which also needs to be divided into an equal number of individual goings. Building Regulations restrict the rise and going for domestic stairs. This is limited to a maximum rise of 220mm and a minimum going of 220mm. All the rises and goings should remain constant throughout any flight of stairs.
- The maximum pitch of a domestic stair is 42°.
- The minimum headroom is 2m, measured vertically from pitch line.
- Obstructions on and around the staircase should be avoided. The

width and length of landings (an unobstructed space at the top and bottom of the stairs) must be at least the width of the staircase. Doors opening onto a landing must provide a minimum of 400mm space between the door swing and the stair.

- Lighting should be considered. Ideally, stairs should be lit by two sources of light (to avoid excessive shadow and provide back-up in the event of one source failing).

Side and landing protection

- Stairs of less than a metre wide should have at least one handrail, and should have one on either side if more than a metre wide or if they are open on either side. Handrails should be parallel to the pitch of the stairs and at a height of between 900mm and 1000mm above the pitch line.
- Stairs, landings and stairwells with a drop of more than 600mm ie the vast majority, should be guarded. This can be formed in a variety of ways; walls, handrails and balusters, panelled construction etc. The construction of the guarding:
  o Should not readily enable children to climb it.
  o Should be strong enough to withstand a significant horizontal impact.
  o Should the guarding have openings then they must be small enough so that a 100mm sphere cannot pass through the opening.
  o Guarding should rise to a minimum height of between 900mm and 1000mm above floor level/the pitch line.

The stairs as a means of escape

The stairs are critical in terms of providing a means of escape in the case of fire. A number of issues require consideration:
- An unobstructed means of escape is vital: how many doorways open onto the stairwell?
- Effective fire resistance of the elements of the building (wall structure, finishes, plasterwork, floors and floor finishes, doors etc) around the staircase: Are there highly combustible materials, ie timber wall panelling, around the staircase? What fire resistance do the doors and understairs plasterwork provide? What about the staircase itself?
- An identification of the level of fire risks in areas adjacent to the staircase: where are the greatest areas of fire risk (kitchens, boilers, rooms with used fireplaces) in relation to the staircase?

Stairs

Treads and risers are jointed together at the front edge but not glued. In order to avoid movement at this vulnerable point, a series of timber blocks are glued to the underside of the joint between tread and riser. Not including these, or their failure over time, is responsible for much of the creaking in stairs. The bottom of the risers are screwed or nailed to the backs of the treads.

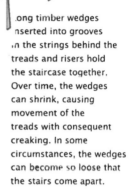

Long timber wedges inserted into grooves in the strings behind the treads and risers hold the staircase together. Over time, the wedges can shrink, causing movement of the treads with consequent creaking. In some circumstances, the wedges can become so loose that the stairs come apart.

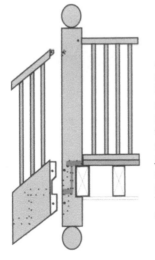

A newel post is frequently used to terminate one of the strings at both high and low level. The string is usually jointed into the newel using a pegged mortice and tenon joint, with the newel itself notched over and fixed back to a trimmer joist. The newels provide a good point to fix handrails and associated guarding for the stairs and the stairwell. The handrails, like the strings are usually morticed into the newel.

Stairs – key design safety issues

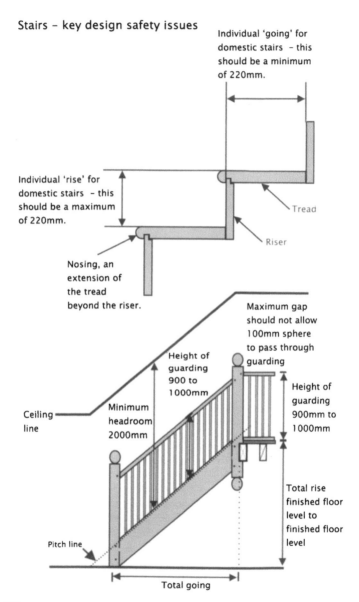

Individual 'going' for domestic stairs – this should be a minimum of 220mm.

Individual 'rise' for domestic stairs – this should be a maximum of 220mm.

Nosing, an extension of the tread beyond the riser.

Tread

Riser

Maximum gap should not allow 100mm sphere to pass through guarding

Height of guarding 900 to 1000mm

Minimum headroom 2000mm

Ceiling line

Height of guarding 900mm to 1000mm

Total rise finished floor level to finished floor level

Pitch line

Total going

Design

The simplest staircase is a straight flight. To save space, landings part-way between floors can enable a change in direction of travel. Such a change can also be achieved using winders: these are tapered treads. They are not as safe as landings because of the narrowness of the tread at one end, however they are popular as they enable stairs to be designed for spaces with restricted going.

Domestic staircases were generally constructed using softwood timbers and consist of a series of horizontal treads, supported 20mm or so back from their front edge by vertical risers. The portion of the tread that extends beyond the riser is known as the nosing. The purpose of the risers is to support the front edge of the tread as it is this point where the load of those using the staircase occurs. The top of each riser is set into a groove on the underside of the upper treads, the bottom of the riser is fixed to the back edge of the lower tread. The width of both the tread and riser is dependent on the specific circumstances, however, the thicknesses are historically consistent, a minimum thickness for a tread is 25mm finished size, this is increased to a minimum of 32mm in 'open stairs' (where there are no risers). In such staircases, the regulations require a minimum nosing width of 16mm. Risers vary from 12–25mm in thickness.

Both the treads and the risers are supported by a pair of beams, known as strings. In a straight flight of stairs the strings run from the lower floor to the upper floor. The most common means of securing the treads and risers to the strings is via trenches (or grooves) cut into the strings into which the treads and risers are secured by long timber wedges. The wedges compress the treads and risers against the internal edge of the grooves in the string. Frequently small blocks of timber are glued to the underside of the staircase where the riser and tread meet. This reduces the degree of deformation and movement at this vulnerable point.

It is common to joint the ends of the string into a vertical post known as 'newel'. The newels provide the means to fix handrails and associated balusters, or other forms of guarding, for both the staircase itself and the upper stairwell or opening.

Common problems with stairs

Creaking and movement of treads and risers as the stair is used is the most commonly encountered defect. The most common cause of creaking is failure or lack of glue blocks at the joint between tread and risers; this allows movement, hence friction and creaking between these components. Shrinkage and failure of the timber wedges securing treads and risers to the strings can also cause movement. Over time, the wedges can work loose to the point where the staircase can start to split apart, with the treads and risers parting company from the strings.

Movement of a staircase may also be caused by the failure of the string fixing. It is common for at least one of the strings to be placed against and fixed to a masonry wall. Over time, these fixings can become ineffective, allowing the string to move as the stair is used. Movement is more common in winders as the vulnerable front edge of such treads are longer and more difficult to support effectively than standard treads. Over time, winders are more likely to weaken than standard treads.

Inadequate sizing of components can enable excessive movement and, more seriously, can allow for structural failure through splitting or shear failure of components. Splitting in inadequately sized components, such as strings and treads, is not uncommon and can be difficult to remedy. Inadequately sized guarding components (ie handrails, newels, balusters etc) is potentially serious and care should be taken in inspecting all guarding components for stability and strength. In addition, inspection should also ensure the appropriate spacing of guarding components.

Other issues that should be considered while undertaking any inspection of the stairs include any obstructions, the quality and quantity of natural and artificial lighting, and the location of glass panels around the stairwell (ie glazed doors).

CHECKLIST

Areas to inspect	Comments/problems

Guarding components

Check for degree of compliance with current of regulation. Specifically check:
- Handrail heights on stairs, at landings and stairwell openings.
- Strength and stability of balusters/rails/panelling.
- Spacing of balusters/rails/panelling.

Safety in-use of the staircase

Check for:
- Excessive steepness (42° is the current regulation maximum).
- Equally dimensioned rises and going throughout the stair (except in tapered treads).
- Minimum headroom while climbing and descending stairs.
- Handrails without interruption (discontinuity can cause problems).
- Levels of natural and artificial lighting.

Strength and stability of the staircase itself

Check for:
- Integrity of strings, treads and risers.
- Significant movement and creaking when climbing and descending the staircase.
- Gaps between tops of string and wall plaster.
- Cracks in components.
- Excessive wear, particularly the treads.
- Missing glue blocks and wedges if there is access to the underside of the staircase.

Staircase as a means of escape

Consider:
- The construction surrounding the staircase, including the doors opening onto stair opening.
- Proximity of fire loading around the staircase, ie where is the kitchen in relation to the stairs, where is the boiler etc.

DAMP

INTRODUCTION

Distinguishing condensation, rising dampness and penetrating dampness is not always easy. An incorrct diagnosis will usually lead to inappropriate repair work; work which not only fails to cure the problem, but may even make it worse. Fortunately, a rudimentary understanding of condensation and dampness, backed up with some fairly simple site tests, can minimise the risks of misdiagnosis and the costs of unnecessary consequential repairs. There must be thousands of houses in the UK that have been re-rendered, partly replastered, or that have had replacement DPCs, all because of some inadequate and incomplete inspection carried out by a surveyor – possibly employed by a damp proofing company that has a vested interest in finding dampness.

CONDENSATION

Condensation occurs in almost every house in the UK. In most cases it will be a temporary phenomenon, perhaps just causing a bit of misting in the bathroom after the shower has been used. In more serious cases it can ruin furniture and decorations; it can also be a risk to health.

Air contains water vapour; if the air is warm is can hold substantial amounts. If moist air comes into contact with a cold surface the air next to the point of contact is cooled. If the air is cooled below a particular temperature (called the dew point), the water vapour will condense on the cold surface. Whether or not condensation occurs depends on the amount of water vapour in the air and the temperature of the surfaces in contact with the air. If you pull back the bedroom curatins in the morning and see moisture on the inside of the glazing it is caused by condensation. Water vapour in the air comes into contact with the cold surface of the glass and condenses. It can be stopped by limiting the amount of vapour produced, ventilating rooms to remove moist air, or warming up the various surfaces to prevent the moist air from being cooled.

Condensation on glass (even double glazing) is common and not too much of a problem. However, in some situations condensation can occur on walls, floors and in roof spaces. Some of these situations are shown on the next page.

Condensation – part 1

Condensation can occur in many places. The obvious ones are in bathrooms and kitchens, particularly if they are badly heated or poorly ventilated. It can also occur in roof voids that are not properly ventilated, around the edges of upper-floor ceilings (insulation not pushed into the eaves), on cold water pipes and on toilet cisterns. It can even occur on cold concrete floors. The wall in the photo (top right) shows extensive condensation in a 1980s flat. The occupier heated the flat with a portable gas fire – these produce water vapour as the gas burns (hydrogen combines with oxygen in the air). The picture on the right shows condensation on a kitchen wall (an external wall, 1 brick thick), again in a property with no central heating.

Condensation is very common on single glazing and not uncommon on double glazing.

In the left-hand photo, condensation has actually occurred within the double-glazed unit (on the inside face of the outer pane), and in the right-hand photo it has occurred on the inside of a stone window surround.

The graphics below show where 'cold bridges' can occur around openings. Where walls are insulated, cold bridges can be at substantially lower temperatures than the main part of the wall. Condensation here is easily confused with penetrating dampness.

Modern and 1920s head sections

Jamb section

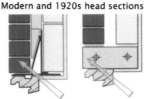

Sill section

Condensation – part 2

What does it look like?

It is sometimes difficult to confirm whether
condensation has taken place because, despite
evidence of its occurrence, it may not be taking
place when a property is inspected. The tests
described on later pages will help to
distinguish it from other damp problems.
Besides the tests, look for its tell-tale signs:

- The damp wall has a 'misty' surface.
- Stains or streaks of water running down a wall, particularly in bathrooms and kitchens.
- Stains or streaks below windows.
- Damp patches with no definite edges.
- Dampness behind wall cupboards, wardrobes etc against external walls.
- Patches of mould – often on pipes and skirtings, not just walls.

Mould may form if humidity levels stay above 70% (see next page) for long
periods. The mould is sustained by organic matter (dirt, grease, food) and
water. The organic matter tends to accumulate on rough surfaces such as
emulsioned walls, embossed or woodchip wallpapers, and tile grout.

Where does the moisture come from?

- showers
- tumble driers (unvented) or drying clothes over radiators
- cooking
- having a bath
- burning bottled gas
- large dogs
- penetrating or rising dampness.

If humidity levels exceed 70% the atmosphere will be damp or 'clammy'.
This can occur during the winter months or during the summer – in the
latter, the warmer air will be capable of holding more moisture. Generally,
45–65%rh provides the most comfortable conditions for everyday living.

Relative humidity

Relative humidity (usually expressed as a percentage) is the amount of water vapour that air contains – relative to the amount it can hold at saturation point (100%rh). If air is warmed, the amount of water vapour it can hold increases; therefore, the %rh falls even though the actual amount of water remains the same. If the air is cooled, the amount of water vapour it can hold reduces; therefore the %rh rises.

Relative humidity in itself cannot be used to quantify the amount of water vapour in the air unless the temperature is taken into account. For example, at different air temperatures, say 10 degrees Celsius and 25 degrees Celsius, an RH of 70% will represent different amounts of water vapour.

Air temperature	RH	Amount of moisture in the air
10 degrees	70%	5.4 grammes/1,000 grammes dry air
25 degrees	70%	13.6 grammes/1,000 grammes dry air

It should be clear that air at the higher temperature (25°) is capable of holding higher quantities of moisture or water vapour. Furthermore, if the air in the first example is heated, the %rh will fall. A rise to 20 degrees Celsius would make the %rh drop to about 40%.

In the first example, condensation (air at 10° and 70%rh) will occur if the air cools to 5° (100%rh). So, if the air is in contact with, say, a cold, single-glazed window, and the temperature on the surface of the glass is 5° or less, condensation will occur. This temperature (in this case 5°) is known as the dew point. The table on the next page shows dew points for a range of humidities and temperatures – it also includes a few more examples.

If we know the air temperature and the relative humidity we can work out the dew point (using the table on the next page). First, we need to work out the surface temperature where we think condensation is taking place. If it is above the dew point, condensation cannot be taking place. If it is at or below dew point, condensation will take place.

Relative humidity can be assessed with a hygrometer. Nowadays, these are usually hand-held electronic instruments with a simple digital read-out. Most modern hygrometers will measure the relative humidity, the air temperature and even the surface temperatures of building materials.

Condensation – part 3

If the internal surface temperature and the relative humidity is known it is easy to work out the dew point using this table. So, for example if the air temperature is 18 degrees Celsius and the relative humidity is 70% the dew point is 12 degrees. In other words, at 12 degrees the air will reach saturation point (100%rh) and condensation will occur. Any surface at or below 12 degrees will suffer from condensation.

Air temp	RH 20%	RH 30%	RH 40%	RH 50%	RH 60%	RH 70%	RH 80%
10				0	2	5	7
11				1	3	6	8
12				2	4	7	9
13			0	3	5	8	10
14			1	4	6	9	11
15			2	4	7	10	12
16			2	6	8	11	13
17			3	6	9	11	13
18		0	4	7	10	12	14
19		1	5	8	11	13	15
20		2	6	9	12	14	16
21		3	7	10	13	15	17
22		4	8	11	14	16	18
23		4	9	12	15	17	19
24	0	5	10	13	16	18	20
25	0	6	10	14	17	19	21

If the internal surface temperature is above dew point but the centre of the wall is below dew point there is a risk that interstitial condensation will occur. The water vapour, therefore, condenses inside the wall rather than on its surface. Interstitial condensation can be a problem with 1 brick thick solid walls, especially if they have been drylined. It is not normally a problem in cavity walls. Drylined solid walls must include a vapour control layer just behind the plasterboard (on the warm side of any insulation). This stops the moist air from reaching the cold surface.

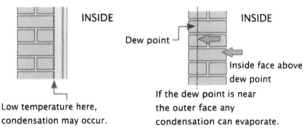

INSIDE

Low temperature here, condensation may occur.

Dew point

INSIDE

Inside face above dew point

If the dew point is near the outer face any condensation can evaporate.

Surface temperatures

Measuring the surface temperature of a wall or window is easy nowadays using modern electronic instruments. But what factors influence surface temperature? There are three:

- the thermal transmittance of the material (U value)
- the outside air temperature
- the inside air temperature.

Well-insulated materials (with low U values) will have high inner surface temperatures and poorly insulated materials (high U values) will have low internal surface temperatures. So, for example, in a house the coldest surface will probably be single glazing. By comparison the internal surface of a modern cavity wall will be very warm because its U value is very low.

Consider two houses – one built in 1930 and one in 2004. The earlier house has single glazing and solid 1 brick thick walls. The modern house has double glazing and insulated cavity walls. If it is 20° Celsius inside and 0° outside then the inner surface temperatures will be similar to those set out below.

	Windows	Walls
1930	10 degrees	15 degrees
2004	15 degrees	19 degrees

It should be clear that the older house is theoretically more at risk from condensation because of its low internal surface temperatures. If, for example, the relative humidity is 70% (remember the air temperature inside is 20°) the dew point temperature is 14°. At this humidity level condensation would be occurring on the glass but not on the wall (it's 1° above dew point). However, if, say, the tumble drier is switched on and the relative humidity increases to 80%, the dew point will rise to 16° and condensation will also occur on the wall.

A checklist for condensation identification is provided after the next sections on Rising and Penetrating Damp but first – just what do we mean by damp?

A definition of damp

The word 'damp' is confusing because nearly all building materials contain some moisture. The moisture content of a timber skirting, for example, will always be greater than 0% and will vary according to the relative humidity of the surrounding air. In humid conditions the moisture content will fall into a state of equilibrium with the air. If the air has a relative humidity of 70% this will be about 16%. The figure below shows the relationship between timber moisture content and relative humidity.

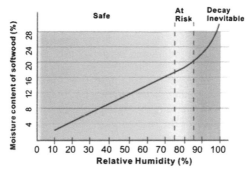

The figure shows that wood with a moisture content of less than 17% is safe from decay. At 17-20% it becomes vulnerable, and above 20% decay is almost inevitable (assuming the timber stays wet for long periods). Modern electronic instruments can accurately assess the moisture content of wood (see page 214 on Moisture meters). But what do we mean by moisture content? Moisture content can be defined as "the amount of water in a material divided by the dry weight of the material". Thus, two materials, one heavy and one light, will have different moisture contents even though they contain the same amount of water.

Moisture contents of other materials

Timber, as demonstrated above, can vary from 4-17%, depending on the relative humidity, and still be 'safe'. Brickwork in the same environment (less than 70%rh) would have a moisture content of about 1-2% and gypsum plaster about 0.5-1%. Brickwork with a moisture content of 3% is damp, at about 10%, or so, it is saturated.

What is dampness?

A dry material can then include some moisture - but how much? We can say that a material is 'air dry' if it is in equilibrium with a relative humidity of less than 70%. So, anything above this can be considered damp.

RISING DAMP

How rising damp occurs

Most building materials are porous; in other words, they contain tiny spaces or pores. Water has a natural inclination to spread along these pores and, the smaller their diameter, the further it will spread. This phenomenon is known as capillary action. Rising dampness in a building is caused, therefore, by water in the subsoil rising up the pores or capillaries of the materials in the wall.

For the last 100 years or so, houses have been constructed with damp proof courses. Materials such as lead, engineering brick, slate, copper, asphalt, bitumen felt, and more recently polythene, have also been used to form a damp proof course. However, there are still thousands of houses from the 19th and 20th centuries without DPCs and, of course, DPCs can fail for a number of reasons. These include:

- deterioration of the original material
- physical damage of the original material
  (eg by building movement)
- bridging of the DPC
- build up of mortar in the base of a cavity wall
- design errors

The wall above left has a DPC about 75mm above the recently resurfaced footpath. It was 150mm above it. The property on the right has a rendered finish that covers the DPC (in line with the bottom of the pebble-dash). The algae growth on the plinth (inset) suggests that the plinth becomes quite wet.

In addition there may be situations where houses without a DPC but with no history of damp problems suddenly exhibit symptoms of rising damp. This is most likely to be due to a change in the water table or serious leakages from drainage or water services.

Chemical analysis

Moisture meters (see page 214) are helpful in providing evidence of rising dampness but they have their limitations, particulalry when used on materials other than wood. In addition, they can be confused by the presence of salts. Definitive evidence can be provided by chemical analysis – usually from samples of wallpaper or plaster. Subsoil naturally contains nitrates and chlorides. These chemicals are drawn up into the wall in solution and are left behind on the face of the wall when water evaporates. They are hygroscopic – in other words thay absorb moisture from the air – one reason why plasterwork usually needs replacing when DPCs are installed or repaired. These hygroscopic salts are most abundant at the peak of rising damp because this is where most evaporation takes place.

Rising dampness is often accompanied by efflorescence, caused by a number of sulfates and carbonates that are always present in building materials. They are not hygroscopic. They can sometimes crystallise and block the pores near the surface of the brickwork, thus driving damp higher up a wall.

Diagnosis

Rising dampness is comparatively rare. Much more common are problems of condensation and penetrating damp. The checklist for this chapter is in the form of tables that provide a systematic approach to the diagnosis of dampness. However, if in doubt call in an expert but remember that many organisations have a commercial interest in finding rising dampness.

Page 215 shows a cross-section through a late Victorian wall and provides some more information on the nature of rising damp.

Damp meters

Wet materials such as brick, plaster and wood conduct electricity. Measuring a material's ability to conduct electricity can help determine its moisture content. The most common form of moisture meter is the conductance meter. These have two metal prongs which should be firmly pressed into the material being tested. The electrical resistance between the two probes can then be measured. Some meters have a digital readout on a scale of 1–100, others have coloured lights (green, yellow and red) representing 'safe', 'borderline' and 'decay inevitable' situations. The coloured light system is particularly useful when trying to assess dampness in materials other than timber. To measure dampness within a wall, not just at its surface, longer probes can be attached to most meters.

Most moisture meters are specifically calibrated for timber. A reading of 18% in a timber skirting means the moisture content of the timber is 18%. However, when used on other building materials such as plaster, brick and concrete, it should be remembered that the % reading is relative. So, for example, a reading of 60% in a plastered wall means the wall contains more moisture than another part of the wall with a reading of 40%. It does not mean that the moisture content of the plaster is 60%. Most gypsum plasters are actually saturated at about 5% or so.

A modern damp meter

Moisture meter readings cannot always be taken at face value – in the wrong hands they can be misleading. Remember that:

- They do not in themselves prove the existence of rising damp or condensation but, used correctly, they can confirm if a material contains moisture (at safe, or unsafe levels etc).
- It is patterns of dampness rather than single spot readings that really help diagnosis – see Checklists at the end of this chapter for some examples.
- Certain salts present in bricks and plaster conduct electricity and can affect meter readings. However if these salts are present in large quantities it does suggest that the material has been wet at some time.
- Metal foil (sometimes found on the back of plasterboard) will give readings of 100%.
- High readings may be due to hygroscopic salts in the plaster. These salts indicate that rising dampness has been a problem at some time – it does not prove it is still occurring.

Rising dampness

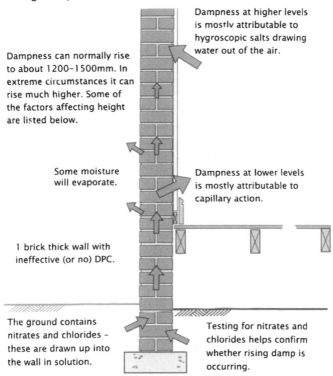

Dampness at higher levels is mostly attributable to hygroscopic salts drawing water out of the air.

Dampness can normally rise to about 1200–1500mm. In extreme circumstances it can rise much higher. Some of the factors affecting height are listed below.

Some moisture will evaporate.

Dampness at lower levels is mostly attributable to capillary action.

1 brick thick wall with ineffective (or no) DPC.

The ground contains nitrates and chlorides – these are drawn up into the wall in solution.

Testing for nitrates and chlorides helps confirm whether rising damp is occurring.

The height to which the damp rises depends on:

- The porosity of the material – in particular the diameter of the pores.
- The amount of water present in the ground and the height of the water table.
- The thickness of the wall – thinner walls are more likely to have high rates of evaporation.
- The nature of any wall finish, for example waterproof render or impervious paint. Such finishes can drive dampness higher.
- The level of heating inside the building. Steady heating encourages evaporation and limits the height of dampness.
- Chemical contamination – some chemicals can crystalise and partly block the pores.

PENETRATING DAMP

How penetrating damp occurs

There are literally hundreds of potential defects that can lead to penetrating dampness. Some of the main culprits are listed below; the pictures on the following two pages show some typical examples. Diagnosing penetrating dampness is sometimes easy; it does not take an expert to find defective guttering or downpipes. However, correct diagnosis can be difficult, especially when the symptoms suggest the problem could also be condensation or rising dampness.

The main causes of penetrating dampness (in walls only) are:

- cracks, usually the result of building movement
- poor detailing, no leadwork over string courses etc
- poor-quality bricks, stone etc
- deterioration of mortar, recessed pointing
- leaks from downpipes, guttering and waste pipes
- leaks from embedded pipes (heating and water services)
- cracked render
- dense walling materials with cracks at lower levels
- copings without DPCs
- copings and sills etc without drips
- inadequate parapet detailing
- lack of chimney flashings and DPCs
- poor window detailing.

Specific problems of cavity walls include:

- wall ties covered in mortar, or upside down,
  or sloping inwards
- cavities full of debris
- missing cavity trays or trays incorrectly positioned
- missing DPCs where cavities are closed
- cavity fill (below ground level) too high
- insulation incorrectly positioned or fixed
- floor joists bridging cavity.

Penetrating dampness – walls generally

Cracked downpipes can produce floods of water in heavy rain. In cast-iron pipes, the back of the pipe (ie the bit facing the wall) tends to rust first because it is difficult to paint. The defective render on the parapet wall (right) is sited immediately above a purlin.

Waste and soil pipes can also leak – their contents are not as wholesome as rainwater! The picture (below right) shows plaster damage below a sash window – damage caused by a cracked subsill.

This render (below left) runs down to the ground and has cracked at DPC level. Water running down the wall has caused extensive staining inside. The owners were convinced it was a DPC problem. In fact the DPC is in sound condition. The problem was resolved by removing the render below the crack and finishing the pebble-dash with a Belcast bead.

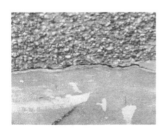

Penetrating dampness – cavity walls

The block of flats shown below was built in the late 1970s. Complaints about damp walls were eventually investigated using a miniature camera inserted into the cavity. The photographs show the problems. The wall ties are covered in mortar and there are broken blocks stuck in various parts of the cavity.

Mortar droppings or sloping ties will both encourage damp pnenetraton.

1970s cavity wall

Missing DPC under coping

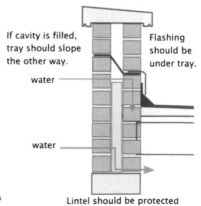

Insulation boards inadequately fixed. They provide a path for damp to cross the cavity.

If cavity is filled, tray should slope the other way.

water

water

Flashing should be under tray.

Lintel should be protected by a cavity tray.

Diagnosis

A damp meter will help to identify the edges of a damp patch and will help plot and monitor its contours. Some typical examples are shown below.

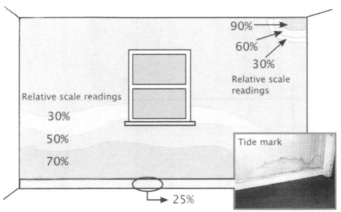

Reading showing moisture content in skirting board.

The damp patch below the window is likely to be rising damp if:

- The height of the top edge does not exceed 1200–1500mm or so.
- There is a sharp drop from high to low readings along the top edge of the damp.
- The skirting gives a high reading.
- Mould growth is probably not present.
- Deep wall probes give high readings.
- Salt analysis shows presence of nitrates and chlorides.

The damp patch below the window is likely to be condensation if:

- The height exceeds 1200–1500mm; it could be the whole wall. Readings may be higher lower down due to temperature difference across wall and due to condensate running down wall.
- Water may wipe onto the hand and there may be obvious streaks of water on the face of the wall.
- There is a gradual drop from high to low readings.
- The skirting gives a low reading.
- Mould growth is usually apparent if the condensation has been there for a time.
- Deep wall probes give low readings.
- Salt analysis is negative.

Distinguishing features

Evidence	Condensation	Rising damp	Penetrating damp
Wall face – feels wet – water wipes onto hand	Very likely, particularly if wall is covered with impervious finish	Unlikely – but decorations may be stained with a tide mark	Depends on cause – face of wall could be very wet
Height above 1500mm	Yes – could be any height	Possible but unusual	Could be any height
Moisture readings at the edges of the dampness	Gradual changes in readings	Sharp changes from wet down to dry	Sharp changes from wet down to dry
Moisture readings in skirtings	Very low readings normal	Readings usually high	High if wall immediately above skirting is wet
Moisture readings in wall (using deep probes)	High surface readings, lower at depth	Readings usually high at any depth	May be higher nearer middle of wall
Mould growth	Very common	Not usual	Not usual – unless very wet
Chlorides and nitrates present	Will not be present	Will usually be present	Will not usually be present
White, efflorescent salts	Only if the wall has been very wet for long periods	Possible	Possible

CHECKLIST

Element	Comments/problems
Preliminaries	If you are examining a suspected problem of dampness check if there are any obvious causes such as bridged DPCs, cracked render or broken downpipes. You should also consider the way in which the property is heated, insulated and not ventilated. Do not forget to consider whether activities inside the house may be contributing to the problem. Portable gas heaters and unvented tumble driers are the main culprits. Similarly, try and assess whether the manifestation of dampness follows any recent changes to the building, for example rendering the external wall.
Contours	If you have access to a damp meter, take a series of readings to try and establish the damp height as well as its contours and edges. Are the changes from damp to dry gradual or sharp? The former suggests condensation.
Probes	Long probes will indicate whether dampness is just a surface problem. In penetrating and rising damp, you would expect to find high readings within the wall.
Skirtings	Check the moisture content of the skirting. Anything over 20% or so suggests the problem is more than just condensation.
Mould growth	Look for signs of mould growth. Check inside cupboards and behind pictures. Mould growth usually suggests condensation.
Staining	Check for tide marks and wallpaper stains. Their height and position can help diagnosis. If the walls are painted plaster, have a good look for signs of efflorescent and hygroscopic salts. The former manifest themselves as white crystals, the latter as damp patches on the wall which may become wetter in humid conditions.

CHECKLIST

Element	Comments/problems
Dew point	If you have the equipment, you should check the relative humidity, the air temperature and the surface temperature of the wall. If the wall is at or below dew point (see table on page 209) condensation is occurring.
Samples	Take some samples of wall paper or plaster. A number of laboratories offer a salt analysis service.
Complex problems	Do not forget that, in some situations, there may be more than one problem occurring at once. For example, rising dampness will lower the temperature of the wall slightly; it will also be adding vapour to the room through evaporation. This may encourage or exacerbate condensation.

ELECTRICAL INSTALLATION

INTRODUCTION

This chapter:
- Describes typical electrical installations.
- Identifies the primary safety issues with electrical installations.
- Explains some of the significant changes that have occurred in both installation, design and equipment.
- Explains common defects that an unqualified person should recognise.
- Identifies more complex defects that require professional inspection and advice.

Electricity is a highly convenient and efficient way of distributing power around houses. However, it is also potentially lethal: each year around 10 people die and 750 are seriously injured through unsafe installations. Unsafe installations are also responsible for 10% of domestic fires. Much of the regulation of electrical installation focuses on ensuring safe installation. The Institute of Electrical Engineers' *Wiring Regulations* (16th edition) outlines the current regulations. Since the start of 2005, the Approved Document, Part P of the Building Regulations, has required that any new installation, and many alterations to an existing installations, must be undertaken by a qualified and certified electrician. Such changes and installations must be notified to Building Control.

Supply

In the past, electrical supply to houses would have been via high-level suspended cables. Nowadays, most houses (with the exception of a few rural areas) are supplied via underground cables: there is less likelihood of damage and less consequential disruption of supply. Electricity is centrally generated and is distributed to substations, which transform the supply to 415 volts. The substations supply the distribution grid with electricity via a cable containing three differently phased supply cables and a neutral return cable. Individual houses connect to one of these three cables – a line or live phase – establishing a standard 230v 'single phase' supply. Each house also has a connection back to the neutral. A single phase supply is adequate for domestic purposes. However, industrial-sized electric motors (1kW or more) generally require connection to all three phases, as well as the neutral return.

Sealed unit, meter and consumer unit

All houses have a sealed unit that terminates the supply cable (known as the service cable). The unit contains a fuse that is connected to the live phase. The fuse is rated at either 60 or 100 amps and is a weak link in the live supply: it will melt or blow if a current greater than the fuse rating reaches it. The neutral cable is not fused and is attached to a solid connection terminal in the sealed unit.

From the sealed unit two individual cables, known as tails, take the live current (red) and neutral (black) through a meter, which measures the amount of electricity used, and onto the consumer unit. Some houses will have a second meter, if the dwelling uses cheaper 'off-peak' electricity, most commonly for electrical storage heaters. The equipment, up to this point, is both owned by, and is the responsibility of, the electrical supplier.

The consumer unit is where the electrical supply is split into a series of individual circuits distributing power (via a series of cables) for different needs around the house. It is both owned by, and is the responsibility of, the house owner. The consumer unit has a switch

Sealed unit, meter and consumer unit with miniature circuit breakers and RCD isolating switch

More recent sealed unit meter, with older wired fuse consumer unit with a simple isolating switch

isolating the entire supply to the building, as well as devices for isolating each circuit individually. In the past wired or cartridge fuses isolated individual circuits but in modern installations miniature circuit breakers (MCBs) have replaced fuses. Individual electrical appliances connected to the circuits usually have their own fuses. The consumer unit also has an earthing terminal for connection to all the individual circuits.

In more recent installations, a Residual Circuit Device (RCD) may be fitted. This constantly monitors the supply current across the live and neutral: if any imbalance is detected, the RCD will shut off supply in micro seconds, avoiding any possibility of electrocution. RCDs can also be fitted to individual circuits where risk of electrocution is high (for example, where they supply the external use of power tools).

Circuits

Multi-core cables leave the consumer unit distributing electrical supply throughout the house. In modern installations there are three types of circuit, each providing electrical power for different purposes:

- **Ring circuits**. These distribute power to sockets, into which individual appliances can be plugged. A multi-core cable (live, neutral and earth) leaves the consumer unit and loops into, and out of, each socket in turn, until it returns to the consumer unit. In this way, the electrical current flows in either direction around a ring, reducing the required cable size. A virtually unlimited number of sockets can be attached to a ring circuit, although the recommendation is that a ring circuit should serve no more that 100m^2 of floor area and total load should not exceed 7.2kW. A ring main will usually have a 30 amp fuse or MCB rating. There is usually a separate ring circuit for each floor in domestic installations. A 'spur' (effectively a radial circuit to another socket or to a fixed fused socket) can be connected off the back of a socket in a ring circuit, though the recommendation is that this should be limited to a single spur per socket. The distribution of sockets should discourage the use of extension leads and overlong appliance flexes, ie there should be sufficient and suitably located sockets. There are regulations regarding the positioning of sockets in relation to sinks, bathrooms etc and in relation to height above floors, worktops etc.

- **Individual radial circuits**. Some 50 or so years ago, radial circuits provided power to individual sockets, which is one of the principal

Typical electrical installation from the 1930s

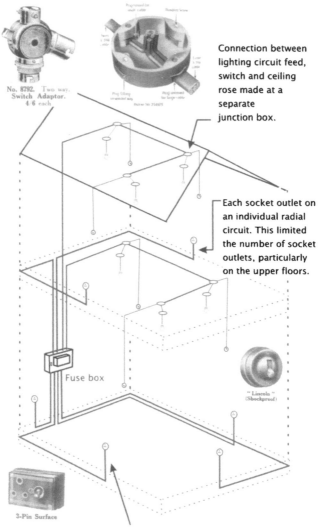

No. 8792. Two way.
Switch Adaptor.
4/6 each

Connection between lighting circuit feed, switch and ceiling rose made at a separate junction box.

Each socket outlet on an individual radial circuit. This limited the number of socket outlets, particularly on the upper floors.

Fuse box

" Lincoln "
(Shockproof)

3-Pin Surface

Individual radial circuits to each socket outlet. More sockets would be found on the ground floor than on the upper floor.

reasons why older installations have a limited number of sockets (this also explains why ring circuits were developed). However, there is still a requirement for radial circuits individually supplying power to high-rated, permanently connected appliances such as cookers, showers and immersion heaters. A twin pole switch is required for such items (where both the live and neutral wires are isolated when switched off). Such circuits have an individual fuse or MCB in the consumer unit protecting the circuit. Early domestic installations (1920s) sometimes only provided lighting circuits. Power for the few electrical appliances available (eg irons and vacuum cleaners) was taken from a light fitting.

- **Lighting circuits.** In modern installations, there will be a separate lighting circuit per floor. If there is a problem with one circuit, the entire house will not be plunged into darkness. Lighting circuits use a relatively low level of power (compared with either of the two previous circuits) and so a radial circuit is normally used (in other words a ring main is not necessary).

  In the past the multi-core cable would run from the consumer unit to a junction box, where a switch and ceiling rose would also be connected. The supply cable would continue from junction box to junction box until the last fitting where the supply cable would terminate. This required a junction box at each light fitting. Nowadays, lighting circuits remain radial but, instead of a separate junction box, the connection between supply and switch in now made within the ceiling rose itself. Lighting circuits have a 5 amp fuse or MCB in the consumer unit and this will support the load of 1200W (ie 12–100 Watt or 20–60 Watt light bulbs).

Cables

Multi-core cables are the norm nowadays. These contain a number of conductors, generally copper wires, which are isolated from one another by being sheathed in PVC. The most common multi-core cable contains three conductors and is known as 'twin core and earth'. Multi-core cables connecting two-way switching require an additional conductor and are the only significant exception; they contain four conductors, including an earth, and are thus known as 'three core and earth'.

The sheathing on the conductors is colour coded to ensure continuity

Typical modern electrical installation

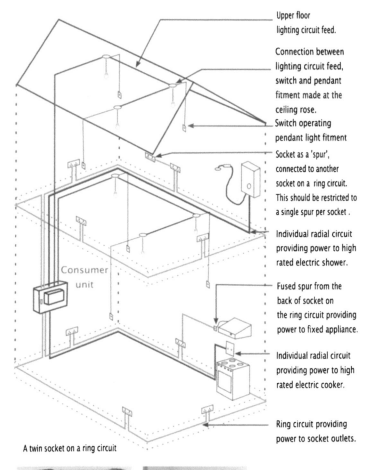

Upper floor
lighting circuit feed.

Connection between
lighting circuit feed,
switch and pendant
fitment made at the
ceiling rose.

Switch operating
pendant light fitment

Socket as a 'spur',
connected to another
socket on a ring circuit.
This should be restricted to
a single spur per socket.

Individual radial circuit
providing power to high
rated electric shower.

Fused spur from the
back of socket on
the ring circuit providing
power to fixed appliance.

Individual radial circuit
providing power to high
rated electric cooker.

Consumer
unit

Ring circuit providing
power to socket outlets.

A twin socket on a ring circuit

Light switch. This is at the
top of a staircase and is
therefore a two-way switch.
The light is operated by switches
on each floor.

of polarity. This coding has recently changed to harmonise coding throughout the European Union. Pre-March 2004, the coding was:
Red=live Black=neutral
The earth conductor is unsheathed. However, where exposed, the earth conductor should be covered in a separate PVC sheath, which is green and yellow.

There is a two-year period of transition between the existing and the new colour coding, however by March 2006 all new wiring (ie changes are not retrospective) should be:
Brown=live Blue=neutral
Earth will be as above.

Whatever the coding the three isolated conductors are themselves contained within a white or grey PVC sheath.

The colour coding of conductors has varied over the years. This can cause problems when extending or altering an existing installation.

The greater the electrical current, the thicker the cable needs to be. This is measured as the cross-sectional area of the copper conductors within the cable itself, for example, a $1.0mm^2$ cable is suitable for lighting circuits, a $2.5mm^2$ for ring circuits and a $10mm^2$ twin core and earth cable for an individual radial circuit for a 13kW cooker. Using undersized cables leads to the conductors overheating. This can damage the insulating material and ultimately can cause a short circuit; potentially lethal if the earthing arrangements are inadequate. Short circuits can also lead to fire.

Since the 1960s, PVC has been the most widely used insulator for both the individual conductors and the sheathing for the cable itself. Before this, a variety of insulating materials were used. From the 1940s, most cables were rubber insulated, and the outer sheath tended to be black. It is not uncommon to find such cables in existing older installations. These cables are beyond their expected safe working life: the extent to which the insulation has deteriorated is dependant on exposure to sunlight, the extent of overloading of the circuit and/or exposure to excessive temperature. The insulation can lose its flexibility, becoming dry and cracked. Replacement at the earliest opportunity is recommended.

Earlier still, cables were lead sheathed, the individual conductors being sheathed in rubber with all the attendant problems outlined above. Unless the lead sheathing is properly earthed, there is a risk of electrocution. Therfore any lead sheathed cables should be regarded as potentially lethal. They are well beyond their safe working life and should be replaced.

Cables distributing power to circuits should not be confused with flexes that are permanently attached to electrical appliances. A flex ends with a plug to connect the appliance to a socket on a ring circuit. The colour coding for flexes is:
Brown=live; Blue=neutral; and Green and Yellow=earth. These colour codings are not affected by recent changes to cable coding.

Conduits and cable protection

Cables distributing power around the house are generally fixed and concealed within the structure of the building. Cables running through floor joists should always be placed in holes drilled through the centre of the depth of the floor joists. If they are in notches on the top of joists, they can be inadvertently damaged.

There are Building Regulations regarding the location of holes and notches in floor joists (see chapter on Upper Floors). Horizontal cable runs should be concealed within floor voids and the cables should be clipped back to the building structure. Where cables run along walls they should run vertically. Generally, vertical cable runs will be concealed within/behind the wall finish and, in order to protect the cables from accidental mechanical damage, these cables should be placed within conduits. Nowadays, these are mostly uPVC although, in the past, metal cover conduits were common. In some instances, no conduits are used at all. These cables are easily damaged (by accident or by rodents) and rewiring becomes difficult. In some cases (often DIY work or cheap rewiring), cables are surface mounted. There are a number of forms of trunking (prefabricated plastic box sections with detachable covers) that can protect and conceal such cables.

Fitments

Fitments have changed little in the last 30 years. The most common material is plastic, although metal fitments are still made. Sockets, switches, ceiling roses and fused connection units are the most obvious parts of the electrical installation; they are also the parts most susceptible to damage and wear. Sockets located too close to the floor often suffer from impact damage and pendant light fittings become brittle because of the heat from light bulbs.

Mounting boxes are fixed to the wall behind switches and sockets. They house the cable bringing power to the outlet. Most mounting boxes are galvanised steel although plastic ones are common within plasterboard partitions. Mounting boxes can become detached from the wall through careless use of the fitting or through poor initial fixing. These are dangerous and require immediate attention.

Prior to the early 1950s, sockets were of two or three round pin format, rather than the square three-pin socket which is universal in the UK today. Until the mid-1960s, fitments (sockets, switches etc) were sometimes mounted on wooden blocks. These can be a fire risk as their design can enable unsheathed conductors to come into contact with a flammable material.

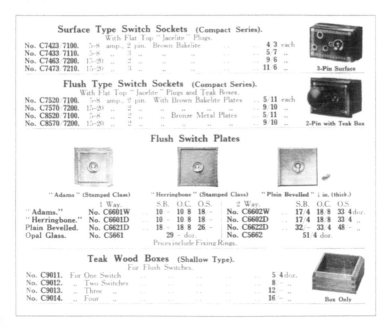

1930s catalogue offering round two and three-pin sockets and teak mounting boxes for metal switches.

It is not uncommon to find that new fitments have been fitted to an existing wiring installation. Clearly the condition and quality of the wiring in such circumstances needs to be checked.

Earthing

Continuous and permanent earthing is a key safety feature of any electrical installation. If a live conductor comes into contact with any metal object, and this object comes into contact with a human, he or she will receive an electric shock. In order to avoid this, a path of least resistance is required which enables current to flow safely through an electrical circuit, rather than through the body. This path of least resistance is provided by the provision of an earthing circuit that connects all the metal parts of an electrical installation and enables any current entering the circuit to safely discharge to earth. If current enters the earthing circuit there is a consequential rise in the current; this will blow a fuse or trip an MCB in the consumer unit. All the connections on the earthing circuit should return back to a terminal block in the consumer unit.

The connection of the earth circuit from the consumer unit to a suitable and safe earth is essential. There are a number of options. In the past, the earthing circuit was connected to a terminal on the coldwater rising main. This steel pipe was a good conductor of electricity and was in direct contact with the ground, thus providing a good earth. However since the 1960s, there has been a concerted programme of replacement of steel water main supply pipes with plastic. Plastic does not conduct electricity and thus an earth circuit attached to a water main is rendered totally ineffective. Since 1966, such an arrangement has been illegal under the Wiring Regulations. Any such installations require immediate replacement.

Currently, there are three acceptable arrangements for a safe and suitable earth:
1. Onto the metal sheathing protection around the service cable itself. This is an older form of earthing arrangement and is gradually being replaced by point 3 below.
2. Onto a metal earthing stake which is driven into the ground outside the dwelling. Generally, such an arrangement is found in certain rural or urban areas where there is overhead supply: such a supply may not have an effective earthing arrangement.
3. Onto an earth terminal block contained within the sealed unit. The earth connection at the sealed unit is connected to the

supply's neutral conductor which, in turn, has a permanent earth connection at the local electricity transformer. This arrangement is known as Protective Multiple Earthing (PME) and is not universally available throughout the UK.

Earth bonding

It is not just the metal components in the electrical installation that require earthing. The current Wiring Regulations require that any metal component that an occupant could touch (at the same time as a faulty electrical appliance) is firmly connected, or 'bonded', to the earth circuit. Generally earth bonding, has a main bonding connection to metal service pipes. This will consist of a special clamp that connects the metal pipes to a main bonding earth cable (sleeved in a green and yellow PVC sleeve) that returns to either the consumer unit or the sealed unit's earth terminal. The clamps should fit tightly around the pipe, be in contact with bare metal and be accessible for inspection. They also have to include a visible warning against removal. The main bonding cable should be a minimum 10mm^2 cross-section copper.

The other metal components, which could be touched at the same time as a faulty appliance, such as sinks, baths, taps, central heating pipes, radiators etc, must be connected to the earthing system via 'supplementary bonding'. Such bonding is achieved using clamps that secure a continuous earthing cable from the metal item to either the earth circuit itself (ie to the earth connection in the mounting box of a socket, switch etc), or to a main metal supply pipe which itself is connected to a main bonding connection. In bathrooms and kitchens, where there are a number of metal items in close proximity that require bonding, a continuous earthing cable can connect from clamp connection to clamp connection and, ultimately, to the earthing circuit or main bonding connection. Again, the clamps must be firmly in contact with the bare metal of the pipe, in an accessible location and must include a written warning.

Bathrooms

The potential for electrocution in bathrooms (wet bodies are less resistant to electrical shock) is so significant that specific rules regarding electrical installations apply. There should be no standard ring circuit socket outlets within the bathroom, light switches should only be string pull type (with no wall-mounting box), pendant light fittings are unacceptable and only low-voltage shaver sockets are acceptable. Earth bonding of all metal components is of great importance in bathrooms. The location of electrical fitments in relation to the water appliances must also be carefully considered.

Inspection issues

A number of the professional associations concerned with electrical installations recommend regular inspections and testing of domestic electrical installations. Such inspections should be carried out by suitably qualified and experienced engineers. The consensus is that, at a minimum, this should occur every 10 years. Typically such inspection should verify:

- The effectiveness of the earthing and bonding arrangements.
- The suitability and serviceability of the consumer unit and fitments.
- The suitability and serviceability of the wiring system.
- The extent of defects, damage, and wear and tear of the above.
- The provision of RCDs for sockets likely to be used externally.

Typical installation faults

There are a number of faults that an unqualified person should be able to recognise:

- Physical damage/wear and tear to electrical fitments. This can indicate a lack of care and maintenance and/or an installation requiring renewal. Any damage to fitments that exposes any wiring, or renders the unit loose, should be isolated from use and professional attention sought.
- Loose, unprotected and bunched cables and cables clearly passing underneath thermal insulation. Cables should be clipped back to the structure, they should run in protective conduits where they pass under wall finishes. Potentially, any cable can overheat. Normally, such heat is able to dissipate, but where cables are bunched and /or where they run under thermal insulation (ie most commonly seen in the roof void at upper floor ceiling level) the build-up of heat can be a potential fire risk.

Unprotected and unfixed cables
covered by loft insulation:
a serious safety risk.

- Scorch marks on sockets and overheating plugs. This can be caused by poor-quality fitments, loose connections at wiring terminals, chronic overloading of the circuit or an incorrect fuse or MCB for the circuit.
- Burning smells (sometimes referred to as a 'marzipan'-type smell) associated with the electrical installation can be caused by:
  - o Loose connections where an arc may occur and where plastic elements of the fitments may be scorched.
  - o Overloaded circuits with the cables overheating and damaging the insulation.
  - o Light fitments with an inappropriate wattage bulb. This frequently results in heat damage to the fitment.
- Frequent tripping of MCB/blowing of fuse on circuits can indicate an incorrect fuse/MCB rating on the circuit, overloading of the circuit or a fault on an accessory or its flex.
- Loss of power at an individual socket can be caused by a wiring failure at the socket, a blown fuse/tripped MCB on the circuit or even the RCD tripping (if there is one). Check the socket by plugging in a different appliance - it could be the fuse in the plug of the original appliance.
- Total loss of power can occur because of mains failure, a blown main fuse or an oversensitive RCD.
- Excessive use of adaptors. Modern ring circuit installations should provide a suitably located and sufficient number of socket outlets. Adaptors can lead to overloading of the circuit and excessive wear to the socket into which they are plugged. Adaptors themselves can become overloaded, overheated and possibly incur scorch marks. Excessive use of adaptors can indicate that there is an old wiring installation even though the sparsely located sockets may be new. Ideally, each item of electrical equipment should have an individual socket.
- Bulbs repeatedly blowing can be caused by mains voltage fluctuations, inappropriate wattage bulbs or cheap bulbs.

Avoid the use of adaptors.

Simple plug-in testers, such as this, can be used to identify correct wiring connections and faulty earthing. These can be useful diagnostic aids but are no substitute for inspection by a qualified and experienced electrician.

CHECKLIST

NB: THIS CHECKLIST SHOULD NOT BE SEEN AS A SUBSTITUTE FOR REGULAR PROFESSIONAL MAINTENANCE INSPECTIONS.

Areas to inspect	Comments/problems
Consumer units	In older consumer units wire fuses are common. Each circuit should have an appropriately rated fuse wire protecting each circuit. Typically these ratings would be: • 5 amp – lighting circuits • 15 amp – radial circuits for an immersion heater • 30 amp – for ring circuits. Check cartridge types, fuse ratings and the rating of more recent MCBs at the consumer unit.
Fitments (sockets, switches etc)	Check for type: any round pin sockets or fitments fixed to wooden backs indicate an aged and potentially dangerous installation that requires professional advice and, probably, replacement. Check for condition: are the fitments cracked or damaged in any way, are the fitments securely fixed, are there any scorch marks or signs of burning?

Areas to inspect	Comments/problems
Use of adaptors and overlong appliance flexes	In a modern installation, there should be sufficient and suitably located socket outlets, avoiding the need for excessively long flexes, extension cables or socket adaptors.
Burning smells or plugs hot to the touch	Check for: • old wiring • frayed flexes • loose connections in socket outlets and plugs • over wattage light bulbs • overloaded circuits.
Cables: Age, type and protection	Are there any exposed cables visible? Cables should be protected by trunking or within conduits. All installations should be wired in cable and not in flex: check the colour coding of the conductors (if accessible) – refer to earlier section for current colour coding. How is the cable sheathed? • Lead, or black rubber sheathed cables are beyond their safe working life and are potentially dangerous. Their existence in an installation indicates the need for rewiring. Seek professional advice. • PVC has been widely used for 40 years, look for cracking and signs of perishing on both the outer sheath and in the sheathing to individual conductors where they are exposed.
Cables overheating	Are too many appliances being used on the circuit? Is the cable undersized? Check roof insulation to see if it covers any exposed cables.

Areas to inspect	Comments/problems
Bathroom	Bathrooms should only have ceiling-mounted string pull switches, ceiling-mounted light fitments and low-voltage two-pin shaver sockets.
Earth bonding	Check for the existence of the earth bonding. Metal pipes, metal sinks and metal baths should have a bonding clamp and green and yellow sheathed earthing connection.
Loss of power at individual sockets	Check for loose connections within the fitment itself, it could also be the fuse of an appliance or it could be the MCB protecting the whole circuit.

PLUMBING AND HEATING

INTRODUCTION

This chapter:
- Describes typical installations for:
  - o cold water
  - o hot water
  - o central heating.
- Explains some of the significant changes that have occurred in both installation design and equipment.
- Identifies the most common defects that an unqualified person can recognise and possibly resolve.

Cold water

Cold water supply is largely taken for granted in the UK. The provision of drinkable safe water piped into houses became more widespread in the late 19th century. Prior to this, supply was a hit and miss affair, with wells, communal standpipes and rainwater collection providing most houses with water. Concerns about public health resulting in legislation, combined with industrial production of pumps, pipes and improved treatment methods, meant that by the turn of the 20th century most dwellings in the UK had a reliable supply of potable cold water plumbed directly to the house.

Cold water is supplied to houses through an underground supply pipe, usually at least 750mm beneath ground level (to avoid freezing). Just inside or outside the boundary there is normally a buried stopcock, which isolates the supply to the house, and, in modern installations, a water meter is located at the same point. The mains pipe enters the house and turns through 90° to rise vertically into the habitable space. At this point, the supply pipe is known as the rising main and it is a regulation that an isolating valve be installed close to the point where the rising main starts at ground floor level. There is also a requirement for a drain cock at this point: these items enable the cold water supply to be isolated and drained down for maintenance.

Cold water installation within houses can take one of two forms:

- **Indirect supply** – where cold water is stored in the house in a cistern. The cistern is generally positioned at high level to provide the force for pushing the water around the house. Indirect systems should have one low-level outlet (generally the kitchen sink), which provides water at mains pressure. Indirect supply:
    - o Is found in older premises or in situations where mains supply is not entirely reliable.
    - o Provides for a store of water, in the form of a cold water cistern located at high level, approximating to 24 hours' supply, should the mains supply fail.
    - o Reduces wear on the components as supply is at a low pressure.
    - o Can be contaminated if the cistern is not suitably detailed; the cistern itself can cause a variety of maintenance problems.

- **Direct supply** – this has no storage cistern, and provides mains pressure water directly to all the cold water appliances. Direct supply:
    - o Is the standard approach in new dwellings where there is a reliable mains supply.
    - o Means there is little chance of contamination as there is no storage cistern and each appliance has fresh, mains-fed cold water.
    - o Causes a higher degree of wear and tear to the installation.

In the past (even as late as the early 1960s) lead pipework was commonly used to distribute both cold and hot water. Nowadays the most popular material for distribution pipework is 15mm and 22mm diameter copper, though there is increasing use of plastic pipes. Until the 1960s cold water storage cisterns were made from galvanised steel but, again, they are now made from plastics.

Cold water installation problems

Cold water supply pipes are occasionally subject to freezing conditions, particularly where they enter the house as part of the rising main, or where they are located within the roof void (supplying, for example, a storage cistern). Cold water pipes are vulnerable to freezing and thus require lagging where they are likely to be exposed to near-zero or sub-zero temperatures.

Cold water supply

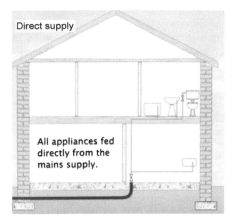

Direct supply

All appliances fed directly from the mains supply.

Where mains pressure is reliable and adequate direct cold water supply is usual. Here, all the cold water taps and fittings in the house are fed directly from the rising main.

A direct cold water system is simpler and cheaper to install than an indirect system and the pressure at the taps is normally higher. A direct supply does not require a cold water tank in the roof space. Cold water tanks are heavy, vulnerable to frost attack, and, in older systems, vulnerable to contamination.

Early water regulations required an indirect water system to have one drinking tap supplied with mains water. However, the law has now changed and requires cold water cisterns to supply potable (fit for drinking) water.

In an indirect system, most fittings and taps are fed by a water tank, or cistern, usually positioned in the roof void. These provide a reserve of water and the lower pressure usually means a quieter system.

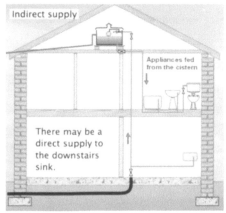

Indirect supply

Appliances fed from the cistern

There may be a direct supply to the downstairs sink.

Mains pressure water exerts significant stress on pipes, pipe joints and bends, and fitments, such as ball valves and taps. Pipes should be suitably supported and secured (particularly plastic pipes, which have a more stringent requirement for support) to the building's fabric. 'Water hammer' is a loud knocking noise created as mains pressure water is forced against the insides of pipes and bends as a ball valve closes (ie at a cold water cistern or at a WC cistern) or at worn tap washers, and can occur in both direct and indirect systems. This can be resolved by fitting an air expansion valve, ensuring pipes are suitably supported or through the use of a pressure-reduction valve on the mains supply.

Wear and tear on washers, pipes and joints is more likely to occur in direct installations. With indirect systems, most problems occur in the storage cistern.

Where there are components of differing metals, for example in older indirect installations, it is common to find galvanised steel cisterns and copper pipes, in this situation bi-metallic corrosion can occur. The water acts as an electrolyte, causing a speeded-up corrosion of the baser metal.

Early hot water systems

Heating water in pans over an open fire or on a range was fairly commonplace in the early 19th century. A more efficient means of producing hot water would have been a 'copper'. This was a vessel, with a direct heat source beneath it, most commonly solid fuel.

Direct and indirect hot water storage

Towards the end of the 19th century, cast-iron back boilers became quite common. These boilers were built-into a flue directly behind an open fire (illustrated right and opposite page from a 1930s builder's catalogue) or a range and took advantage of the significant proportion of heat lost up the flue to directly heat a supply of water that was fed from a high-level cold water cistern. The heated water was stored in a hot water cylinder from where it would be drawn off to the taps. The hot water drawn off from the cylinder is water that had been directly heated in the boiler, thus its name – a direct hot water cylinder. The main

problem with direct hot water cylinders is that as the hot water is drawn off it is replaced with fresh water. Heating water leads to the deposition of salts as they crystallise; the constant replacement of fresh water with dissolved salts can lead to firring–up of both the cylinder and the pipe work.

Back boiler systems are still installed, and are generally part of solid fuel-burning stoves or radiant gas fires. The increasing incidence of electrical installations in houses post-1920 witnessed the development of immersion heaters, electrical elements that directly heat the water in the cylinder.

Direct 'combination' cylinders (sometimes known as Fortic cylinders) have an integral indirect cold water supply and can be used where there is a direct cold water supply, ie in flats where there is no roof void. A much more recent phenomenon is the use of a pressurised direct water cylinder. These are directly fed by the cold water mains and are therefore at mains pressure.

With all hot water production, water becomes lighter and expands as it is heated. This expansion needs to be safely dealt with. Where an indirect cold water system supplies the water to a direct hot water cylinder, this expansion is safely accommodated by a slight increase in the level of water in the cistern. If the controls fail and the water boils it can escape into the cistern through the vent pipe. In a pressurised system, these do not exist; instead, sophisticated controls, safety valves and expansion vessels deal with this potentially dangerous expansion. Therefore, they require frequent preventative maintenance to ensure that they remain fully operational.

In order to avoid the significant problem of firring-up, indirect hot water cylinders became increasingly common through the 1950s and 1960s. In an indirect cylinder, the water drawn-off from the cylinder has not been directly heated. Heat is transferred by heat exchange within the cylinder: water in the cylinder surrounds a coiled copper pipe (a heat exchanger), which contains water that is heated by a heat source (generally a central heating boiler). This 'primary circuit' is also frequently extended to provide heat to radiators. Thus, the big

Direct cylinders

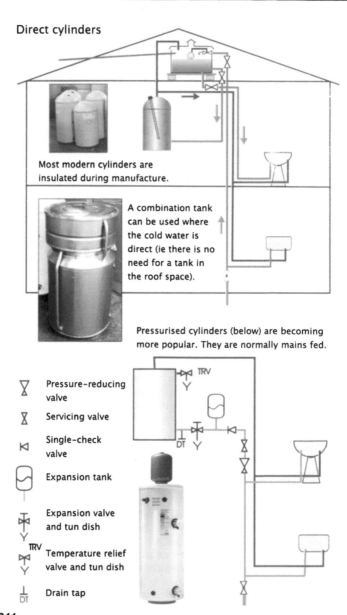

Most modern cylinders are insulated during manufacture.

A combination tank can be used where the cold water is direct (ie there is no need for a tank in the roof space).

Pressurised cylinders (below) are becoming more popular. They are normally mains fed.

⊽	Pressure-reducing valve
⊠	Servicing valve
⊲	Single-check valve
⊖	Expansion tank
⊕	Expansion valve and tun dish
TRV ⊕	Temperature relief valve and tun dish
⊥ DT	Drain tap

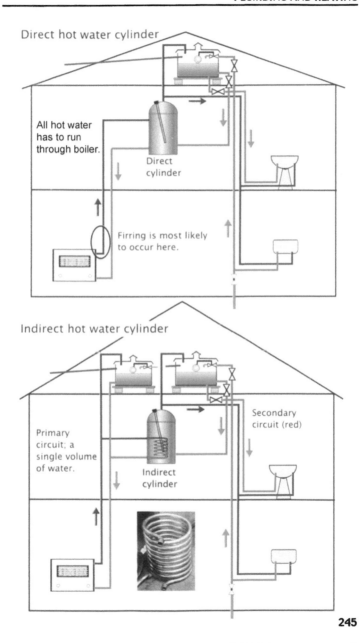

Direct hot water cylinder

All hot water has to run through boiler.

Direct cylinder

Firring is most likely to occur here.

Indirect hot water cylinder

Primary circuit; a single volume of water.

Indirect cylinder

Secondary circuit (red)

advantage of indirect hot water cylinders is that the water in the primary circuit is continuously reheated and recirculated: this water can be chemically treated to avoid firring-up and, even if it does, it will lead to very limited crystallisation compared with direct storage cylinders. Apart from leaks or draining down, the same volume of water can stay in the system for years.

Non–stored hot water systems

Less common than stored systems are instantaneous hot water heaters. As cold water is drawn through such an appliance it is heated as required. Early 'geysers' were gas fired and were only able to produce enough hot water for a single bath or wash basin. Many similar, but more sophisticated, gas appliances are available today. A single-point heater can provide sufficient hot water for a sink without the need for an externally connected flue to remove exhaust gases (although there are time limits to running them continuously). A multi-point will provide sufficient hot water for a number of fittings (say a bath, a basin and a sink) but will require an external flue. Single-point, electrical water heaters are also common. These store a small amount of hot water.

Hot water problems

Long pipework runs between the storage cylinder or a multi-point instantaneous water heater of the taps can result in a significant loss of heat if the pipe is not insulated.

Careful location and planning of pipe arrangements and pipe lagging is required. The Legionella bacteria multiplies in water-based systems. In order to avoid this, it is vital that water is either hot or cold (not tepid). It is recommended that hot water should be stored at a minimum of 60°C, distributed at not less than 50°C and cold water kept at a maximum of 15°C. Where an intermediate temperature is required this should be mixed at the point of use. A problem can occur when cold and hot water supply pipes are in proximity, the hot can heat up the cold and vice versa. This can lead to a situation where the bacteria can multiply. Hot and cold pipe runs should be kept separate or should be insulated where they are in close proximity.

In direct systems (see top diagram on page 245), a common problem is the deposition of salts or 'firring-up' of both the cylinder and/or the pipes. This is most likely in hard water areas, ie where water supplies contain a high level of calcium carbonate deposits. The deposition of salts can reduce the efficiency of the system because it insulates immersion elements, reduces the internal diameter of pipes and, in the worst case, can cause boilers to explode. Water treatment, either by water-softening devices, or by holding the salts in solution via magnetic-type devices, are options in hard water areas.

In direct cylinders heated by older back-boilers, oxygen entering the system can cause oxidisation of the cast-iron back-boiler and deposits of rust may be found in the hot water drawn off from the taps. This is particularly problematic in soft water areas (water that has less calcium deposits and a higher acidic content). Back-boilers also suffer from a lack of effective thermal control and this can cause the hot water in these direct systems to boil, creating expansion, noise and vibration.

Fortic tanks, with their own built-in, indirect cold supply, provide very poor motive force for the hot water and are inadequate for showers unless a shower pump is added.

Space heating

In contrast to today, the provision of space heating in the past tended to occur on a room-by-room basis, in other words fires. These required coal or timber to be physically transported to each fireplace, the fire to be fed as it burned and the ashes to be removed – not very convenient. Open fires were not very efficient either, with a significant percentage of the heat (70% or more) being lost up the chimney.

The early use of 'town gas' (manufactured from coal) was exclusively for lighting rather than heating. By the 20th century, many houses had gas piping. Gas became increasingly used as a means of heating from the 1920s in the form of individual, radiant gas fires. These provided a far cleaner and more convenient form of individual room heating compared with solid fuel. Two examples of typical radiant gas fires from the 1930s are illustrated overleaf.

Natural gas, initially from North Sea supplies, replaced town gas during the 1960s. Mains gas is the most popular fuel for domestic heating systems and this has been the case for the past 40 years or so. The non-mains alternatives are liquid petroleum gas and fuel oil. Both require an individual storage tank and both are more expensive and less convenient than using mains gas. There are stringent regulations regarding the location and detailing of storage tanks.

No. 1403.

No. 1404.
With Boiling Burner.

Individual electric heaters became increasingly popular in the latter half of the 20th century: these had a distinct advantage over gas fires in that they did not require a flue for the removal of exhaust gases. Electric fires could be positioned anywhere in the room. After the early 1970s fuel crisis, it was recognised that a far more efficient and effective form of electrical space heating was required. The most frequently adopted system was the electric storage heater. Storage heaters use off-peak electricity (cheaper tariff electricity produced during the night – hence the term 'night storage heating'). Electric current flows through a heating element that is surrounded by bricks; these store the heat produced. The heat is released slowly during the next day. More recent heaters have fans, dampers and high levels of insulation, making them far more efficient and flexible than their earlier counterparts. Typical older electric storage radiators are illustrated below. The photo (page 249) shows the exposed heating elements – the front cover and most of the bricks have been removed.

Electric storage heaters are still fairly common, and are frequently installed in situations where there is no gas supply available. Such an installation will, obviously, require a separate system for heating water.

By the late 1950s there was a recognition that a far more efficient way to produce heat was to have centralised heat production (for both hot water and space heating) and by the 1970s houses were increasingly being fitted with 'central heating'. The most common form of

space heating in the UK is a gas-fired central heating system. These are 'wet' systems, where the heat exchanger in the cylinder and the radiators are heated by hot water. There are a number of different systems but the two most likely to be encountered are:

- the vented indirect storage system
- the non-vented direct non-stored system.

Vented systems

These systems are becoming less popular, but currently remain the most commonly encountered. They typically comprise:

- **A boiler** – which produces heat, distributed via a primary circuit. All boilers require a flue, most will have a fan-assisted balanced flue that removes exhaust while at the same time providing air to enable the fuel to burn.
- **A feed and expansion tank** – which feeds the boiler and ensures that the primary circuit is always full. The tank also allows for water expansion in the primary circuit and accommodates vented water if the water boils. It is a relatively small tank located at high level (ie within the roof void).
- **An indirect hot water cylinder** – through which the primary circuit heats indirectly supplied cold water. The cistern feeding cold water into the cylinder also provides the opportunity to accommodate any expansion in the cylinder.
- **Various controls**, including:
  - o **A pump**: which provides the motive force for the primary circuit.
  - o **Thermostats**: a variety of temperature operated switches, which control the boiler, hot water cylinder, individual radiators or the whole heating system.
  - o **Motorised valves**: these are operated by the various thermostats in the system and direct the flow of the primary circuit to the radiators, to the hot water cylinder, to both or to stop the flow to both.
  - o **Programmer**: essentially, this is a timer switch that can be

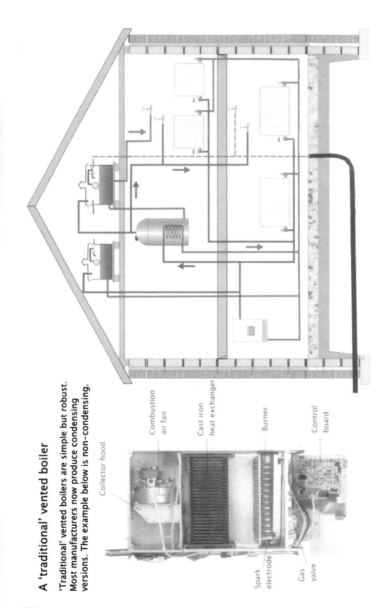

A 'traditional' vented boiler

'Traditional' vented boilers are simple but robust. Most manufacturers now produce condensing versions. The example below is non-condensing.

Collector hood

Combustion air fan

Cast iron heat exchanger

Burner

Control board

Spark electrode

Gas valve

programmed in advance to switch the heating on or off
and/or hot water system at predetermined times. The unit
also co-ordinates data from the boiler, cylinder and room
thermostats and operates pumps and motorised valves.

o **Energy/boiler managers**: unlike the above, this device is
optional and is not always found in every system. There
are a variety of possibilities – some devices simply govern
the boiler to ensure the most efficient use of fuel, others
do this through measurement of external temperature.
There is much scepticism in the trade as to whether or not
such devices are cost-effective for domestic installations.

• A **distribution network**:

Usually copper, although plastic is becoming more common.

o Pipework from the boiler to the cylinder and back
to the boiler.

o Pipework from the boiler to each radiator and back
to the boiler.

• **Heat emitters**: steel panel 'radiators' (in reality, they emit
heat by convection).

These systems require careful planning in terms of the relative
location of the boiler, hot water cylinder, feed and expansion tanks
and cold water cisterns and radiators if an effective and efficient
space-heating system is to be installed. A typical layout is shown on
page 250. Page 252 shows typical controls.

Non–vented systems

A non-vented system has far fewer items of equipment than a vented
one but requires a more sophisticated boiler. The system (as shown
on page 254) does not have any provision for stored hot water and
has, therefore, no cylinder – the boiler acts as an instantaneous gas-
fired water heater as well as providing the heat for the space-heating
circuit. This type of boiler is commonly known as a combination
boiler. The boiler is directly fed from the mains, so there is no need
for a cold water cistern. These boilers were originally developed in
France and Germany, where water supply tends to be 'direct' and
where many people live in flats (ie with limited space for feed and
expansion tanks).

Thermostatic Radiator Valve (TRV): enables room-by-room temperature control (by closing down radiators).

Room thermostat: controls the space heating by switching off boiler when a set temperature is reached.

Cylinder thermostat: regulates the flow of primary water through the heat exchanger.

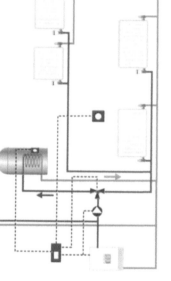

Typical controls on a vented central heating system

Programmer: a 'timer' and electronic co-ordinator of signals from thermostats.

Pump: fitted at an appropriate point on the flow pipe of the primary circuit.

Motorised valves: switches direction of the primary circuit.

A combination boiler:
- Has to cope with relatively high pressure as well as the inevitable expansion of heated water in the primary circuit. This is done via a series of safety devices, such as an expansion vessel and pressure-reduction valves.
- Contains an integral pump as well as thermostats to control the temperature of the heating and hot water circuits.
- Has a fan-assisted flue.

Typically, the installation will have a separate programmer. Air temperature control is established by thermostatic radiator valves that are attached to the flow into each individual radiator.

These systems offer a number of advantages compared with vented systems: they require less planning and have fewer parts and connections. They can be more economic in fuel consumption as they heat hot water only as and when required. However, they rely on a complex boiler, which probably has a lower life span compared with the more robust and simpler boiler of the vented system. Frequent preventative maintenance of such boilers is essential.

Gas-fired systems generally must be installed by a CORGI-registered installer, and if the house is tenanted there is a statutory requirement that all gas-burning installations are subject to an annual gas safety check. It makes good sense that a suitable experienced/qualified engineer should carry out such a check regardless of the tenure.

Since April 2005 Part L of the Building Regulations requires that all new or replacement boilers installed in England and Wales must be 'High Efficiency' boilers; these are mostly condensing boilers. A condensing boiler is more fuel efficient as it recycles latent heat in the exhaust gases, which, in conventional boilers, are allowed to escape through the flue into the atmosphere. However, they are more expensive and, by some accounts, less reliable.

Space–heating problems

Many of the problems that occur with 'wet' heating installations require expert intervention. This section explains some of the more common problems, a few of which are relatively easy to identify – they may, of course, not be so easy to remedy. If in doubt, seek appropriately qualified advice.

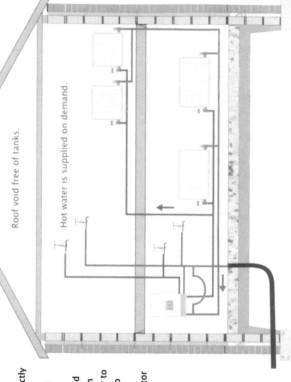

Roof void free of tanks.

Hot water is supplied on demand.

A combination boiler

An installation heated by a combination boiler does not require a cold water storage tank or a hot water cylinder. The heating system is topped up directly via a filling 'loop' fed directly from the mains. The expanding water is accommodated by an expansion vessel fitted inside the boiler (see below). Hot water for washing etc. Is heated instantaneously as, and when, required. A typical installation relies on TRVs and the temperature knob on the boiler to control air temperature. A better method is to include a programmable room thermostat in, say, the lounge. In this case the lounge radiator should not include a TRV.

Wet central heating systems

Boiler problems: Most modern boilers have a sealed combustion chamber and the integrity of the seal is vital to avoid poisonous exhaust gases entering the dwelling. In most modern boilers, a sensor will shut down the boiler if this seal fails. In some older boilers, the room housing the boiler should have adequate ventilation in case of seal failure.

Most modern boilers will have a balanced flue, this requires careful siting on an external elevation away from plastic guttering and downpipes, opening windows and air bricks etc. Such restrictions are less onerous when the balanced flue is also fan assisted. Older boilers, without a balanced flue, draw oxygen for combustion from the room where the boiler is located (the space should not be a bedroom). Clearly, good ventilation for such a space is vital.

Modern boilers will have a thermostat that switches the boiler off if excessive water temperatures are sensed; this vital fail-safe is to protect the occupant and the boiler. This thermostat should be set between 70°C and 80°C, any hotter than this and the radiators would be dangerously hot. The control for this thermostat is usually found within the boiler casing.

Fuel/air mix is critical for efficient and safe operation. The flame burning in the boiler should be blue/purple. If there are any signs of a yellow flame this indicates an unclean, inefficient burn, which will lead to deposits in the combustion chamber and the flue – a potentially dangerous condition whereby an engineer should be consulted.

A noisy boiler can be caused by a build-up of sludge or scale within the boiler's heat exchanger. However, many of the apparent problems affecting boilers are, in fact, created by other components, in particular problems induced by faulty controls. In the vented indirect storage system, the components are separate and this makes inspection, diagnosis and intervention relatively easy. Common problems include:

Pumps:
- Inappropriately located.
- If set too fast can cause noise and vibration to surrounding parts of the installation.

If the radiators upstairs are cold it may mean the feed and expansion cistern tank in the roof has run dry or the boiler pressure (combi) is low.

If the top of the radiator is cold it usually means the radiator needs bleeding.

If the radiators nearest the boiler tend to be the hottest they may not have been balanced properly. Balancing is done by adjusting the lockshield valves (below). Radiators on a one pipe system are always difficult to balance.

Thermostatic radiator valves can sometimes stick – usually in the closed position. This is usually discovered when the heating is first turned on at the end of the summer. The outer casing can be removed and the sliding pin freed with a pair of pliers.

If the bottom of the radiator is cold it may mean the radiator is blocked with rust or sludge.

If all the downstairs radiators are cold the pump may not be working properly.

- Contain bearings that can become worn over time leading to noisy operation, which usually signifies the need to replace the pump.

Room thermostats:
- Inappropriately located, eg above a radiator or in an area where there are likely to be extremes of temperature.
- Used as the sole control for regulating the boiler.
- Being set high to enable speedy heating-up of the space.

Boiler thermostats:
- Set too hot.
- Set too hot for seasonal temperatures.

Cylinder thermostats:
- Need to be set to a minimum of 60°C to avoid the spread of legionella bacteria.
- If set too high can create inefficient fuel use and will produce scolding hot water.

Thermostatic radiator valves:
- Can stick at a particular setting.
- To be effective, should be located on the flow into the radiator rather than on the return.
- Can be difficult to understand without the aid of the technical data or a simple explanation from the installer.

Motorised valves:
- Can suffer if pipework becomes furred up: they also have a limited life and will require replacement over time.

Radiators
- Can be poorly sized for a given location and thus can lead to an inefficient installation (see previous page for more information on radiators).
- Most are steel and, over time, particularly if subject to frequent airlock, can suffer from internal corrosion, rusting and ultimately leaks. In systems with copper, pipes can also enable and encourage bi-metallic corrosion.

Common problems are shown on page 256.

Distribution pipes:
- Can suffer airlocks, the symptoms include blockage, cool radiators, intermittent flow. Radiators have a bleed valve that can be used to remove trapped air. If the radiators require frequent bleeding, this suggests a more significant problem.
- Can cause knocking and other noises when they are poorly secured within the building's structure. This occurs most commonly as the pipes heat up and expand as the system is switched on.
- Can be damaged by impact if located close to the floor in areas of high traffic. Microbore (a softer flexible 10mm diameter copper system popular in the 1990s) can be easily damaged.
- Can suffer leaks at joints: if they are compression fittings they can be tightened, if they are soldered joints they will require remaking after having drained the system down – wet pipes cannot be soldered.

CHECKLIST

NB: THIS CHECKLIST SHOULD NOT BE SEEN AS A SUBSTITUTE FOR REGULAR PROFESSIONAL MAINTENANCE INSPECTIONS.

Areas to inspect	Comments/problems
Location of rising main and gas supply to dwelling	These should be known and be easily accessible in case of emergency/requirement to shut down installation. Check for: • Location and presence of stopcock and drain taps on rising main. This is frequently located under the kitchen sink. • Location and presence of gate valve for mains gas supply (usually located adjacent to gas meter).
Cold water supply	**Pipework.** Copper is the most common material for pipework. Plastic is increasingly found. It requires more support than copper as it has a lower tensile strength. Plastic (sometimes available in long coils) may have far fewer joints than copper. Sagging pipes can loosen 'push-fit' joints. **NB:** Lead pipework is no longer acceptable and should be replaced. Check for: • Leaks and integrity at joints. NB: In bathrooms, kitchens and other areas of high humidity cold water pipes sometimes show signs of condensation, which is sometimes mistaken for leakage. • Support and restraint. • Impact damage. • Adequate separation of hot and cold pipes. • Appropriate lagging/insulation in vulnerable areas (where exposed to zero/sub-zero temperatures – lower floor voids, within insulated roof voids etc). • Check the operation of cold water taps – is the pressure adequate, do the taps drip? Is there any discoloration of the supply? • Identify whether the supply is direct or indirect.

Areas to inspect	Comments/problems
Cold water storage cisterns (indirect or direct supply with a vented hot water cylinder)	Check for: • Leaks around the cistern itself (particularly if cistern is galvanised steel). Leaks are also found at inlets and outlets to the cistern. • Integrity of the lid, essential if the cistern is not to become contaminated. • Integrity of the insulation around sides and top of cistern but not underneath. Include lagging (insulation to pipes). • Overflow/warning pipe must be fitted, insulated and should discharge to the outside. • Integrity of ball float. • Operation of valve and correct water level. • Structural support for the tank is suitable and there are no signs of structural distress. • Appropriate location and security of any expansion/vent pipe from any hot water cylinder. • There should normally be separate tanks for a vented boiler and a vented hot water cylinder. (An indirect cold water system will share a cistern with the feed for the hot water cylinder.)
Hot water cylinders	Identify whether the cylinder is: • Direct or indirect. • Vented or pressurised. Check for: • Primary material: the vast majority will be copper, which has a high degree of resistance to corrosion. It is still worth examining the outlets (top for the hot water distribution pipe/expansion pipe, lower inlet for cold water supply) and inlet/outlet for the indirect coil, for signs of leakage. • Galvanised steel cylinders may still be encountered: these are far more likely to suffer from corrosion-based leakage. • Insulation: cylinders should be insulated. Some cylinders will be provided with a foamed, sprayed-on insulation at the factory, this may reduce the extent of inspection for leaks.

Areas to inspect	Comments/problems
	• Insulation 'jackets' may be found on older cylinders; they can be loosened for inspection purposes. • Leaks around any of the immersion element(s). Check that the wiring is not frayed or damaged in any way. • Support: cylinders are heavy and require appropriate structural support. Pressurised cylinders will require greater vigilance and more regular inspection/servicing. Particularly look for signs of leakage to expansion vessel and other associated safety equipment.
Identify whether boiler is vented or pressurised	• The presence of a separate feed and expansion tank within the roof void and a hot water cylinder (with two outlets in its side for the boiler flow and return) will indicate a vented boiler. • The lack of a separate feed and expansion tank in the roof void and no hot water cylinder will suggest a non-vented combination boiler system. • If there are no radiators, the boiler is likely to be a multi-point water heater.
System generally	Whether the boiler is vented or pressurised (combination) you can check, albeit at an elementary level, whether or not it works. Manually switch on the boiler or alter the time clock. It should ignite immediately assuming the thermostats are correctly set. Check the colour of the flame: it should be blue purple, with no sign of yellow. If this is encountered, switch off the boiler and contact a heating engineer immediately. The boiler should operate without excess noise and be free from leaks, rattles etc. After a few minutes, the radiators (and any cylinder) should start to warm up. Check the radiators for leaks and temperature If you have the time, you can also check whether the thermostats work properly. Turning the room and cylinder thermostats down, for example, should switch off the boiler.

Areas to inspect	Comments/problems
Heating pipework generally	Check for leaks etc as explained at beginning of the checklist. Thin, copper micro-bore pipework (10mm) is prone to squashing and blockages. Steel pipes can be found in older installations; look for signs of rusting. In a combination boiler, check the pressure of the system (there is usually a pressure gauge on the front cover). A low reading suggests the system needs topping up (usually a result of overzealous radiator bleeding or leaks).
Feed and expansion cistern	**NB:** Only found in vented space heating systems. Check for all the points raised in cold water storage cisterns above. In addition, check that the water has been treated with chemical additive to limit firring.
Flues for gas burning appliances	Check for: • Location externally in relation to opening windows, air bricks and combustable/heat-affected components (ie plastic guttering overhanging eaves). • Effective operation while boiler is operating. • Condensing boilers should 'plume' – this is not a defect.

DRAINAGE

INTRODUCTION

This chapter:
- Describes typical installations for:
  - above and below-ground foul and waste drainage
  - rainwater drainage.
- Identifies the most commonly encountered defects in the above systems.

Above-ground drainage

Pre-1950, above-ground drainage systems are likely to be two-pipe systems and are likely to be fixed to the external surface of the building. Two-pipe systems are distinguished by having two separate vertical pipes (or stacks): the 'soil and vent stack', and the 'waste stack'. The soil and vent stack takes discharge from WCs. This stack is directly connected to the underground drainage system and it usually extends up beyond the roof line (to avoid foul smells entering the house) to form a vent pipe that allows air to freely enter and leave the drainage system – hence the term soil and vent stack. An integral trap (the 'hidden bend') in the WC prevents any foul smells from the drain entering the building. See next page for more information on traps.

The waste stack takes effluent from sinks, baths, showers and other domestic appliances, it also takes the discharge from the rainwater drainage systems. This stack usually terminates in an enlarged open end known as a hopper head' where the various waste branches and rainwater downpipes discharge. The open hopper head often gets blocked or allows material to enter the drain itself, again causing blockages. The hopper head, being open, requires a trap at the base of the attack pipe to prevent gases in the underground pipes from escaping.

Kitchen sinks often discharged into a separate trapped gully rather than into the waste stack. Baths, sinks and basins all had traps just under an appliance's outlet. Typically these would be 'P' or 'S' shaped traps or, more recently, bottle traps. Traps forming part of the branch

Traps and siphonage

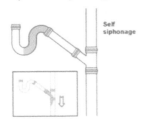

Self
siphonage

Induced
siphonage

Self-siphonage occurs where water running along a branch 'sucks out'
the trap. Induced siphonage occurs where water in a discharge stack
'sucks out' the trap in a branch pipe.

Siphonage can be prevented
by adding vent pipes or by the
use of self-sealing traps.

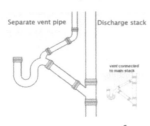

Separate vent pipe

Discharge stack

vent connected
to main stack

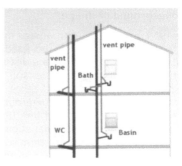

vent pipe

vent
pipe

Bath

WC

Basin

Until the 1950s fittings and pipes
were usually made from cast-iron.

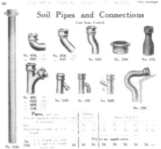

Soil Pipes and Connections

Cast Iron Coated

Single stack system

Self-siphonage is most likely to occur where there are steep gradients or long branches.

Top of stack must be clear of dormer windows etc. It should also include some form of cage to prevent birds nesting.

Wash hand basin

WC

Bath

If fittings are fitted with P traps of less than 75mm depth, or S traps, self-siphonage may occur.

WC discharge must not be allowed to run into waste branch from bath – it should be swept in direction of flow to avoid siphonage.

Kitchen sink

Modern wastes are usually plastic. They can have push-fit, compression ring or glued joints.

Typical single stack system.

WC

Baths, sinks and showers are less susceptible than basins to self-siphonage as the slow discharge of water (due to flat bottom of appliance) helps to re-seal the trap.

If the waste from a basin is connected to a branch, as shown here, siphonage may occur. To prevent it, a vent pipe or self-sealing trap may be necessary.

wastes to sinks, basins, baths etc could be broken by siphonage. This can occur either by 'self-siphonage', where the branch waste runs full bore, sucking out the water in the trap, or by 'induced siphonage', where the vertical stack, into which the branch waste runs, runs full bore.

During the 1950s, following considerable research, the 'single stack' became an acceptable alternative to the two-pipe system, the latter suffering from:

- Blockages caused through detritus entering the hopper heads (which were also difficult to access on high buildings).
- An unacceptable profusion of externally fixed pipework, leading to problems of maintenance (made worse if bathrooms and WCs were added). To reduce the risk of siphonage ventilation pipes were often added but these could virtually double the amount of pipework required.
- High capital cost.

Nowadays, the single-stack system is the most commonly adopted installation. The single-stack system has a single soil and vent pipe into which WCs, as well as sinks, wash-basins, and baths etc discharge. There is no underground trap at the base of the stack and thus the system relies on secure and effective traps located within (as with WCs) or just below the various sanitary fittings. To avoid the potential problem of siphonage, the Building Regulations limit the length, diameter and gradient of waste branches. The stack should enable ventilation of the drain, and thus, as with the soil pipe in the two-pipe system, is generally extended above the roof line and is known as a soil and vent pipe (or stack). In a restricted number of situations, the soil and vent stack can terminate within a building; in these circumstances, the pipe must be fitted with an air admittance valve.

For most houses, it is normal and acceptable to have externally fixed stack pipes: they can be located within the dwelling (engendering less mechanical and exposure-related damage, but requiring suitable access arrangements). For dwellings over three storeys, the pipe work system must be contained within the dwelling.

Rainwater goods

In most roofs, where the rainwater discharges from the roof at the eaves, a gutter collects the rainwater, which then flows into appropriately located downpipes. The effective roof area and likely rainfall pattern should be used to determine the size of the gutter, and the size and frequency of downpipes. There may also be aesthetic considerations in regard to selection.

In some older roofs, most notably where parapet walls hide the eaves, a different, more complex gutter is required. In these circumstances, gutters are formed with a waterproofing membrane (traditionally lead) laid over timber boards.

These parapet gutters usually discharge through the parapet wall itself, via a lead-lined scupper, into a downpipe. In some situations, most notably Georgian and Regency properties, the gutter that discharges through a secret, gutter which runs through the roof void itself, to a downpipe that is also hidden within the building. Secret gutters are notorious for blockage and subsequent internal flooding.

Below–ground drainage

In new installations the Building Regulations require separate underground drainage systems within the curtilege of the property and within a new development. This involves two separate systems of underground pipes: one for the rain and surface water collection, the other for soil and waste effluent. These separate systems might then discharge into separate sewerage systems (where they exist – they are not universal in the UK) or the separate systems may discharge into a combined underground sewerage system.

Older installations, particularly those with a two-pipe, above-ground drainage system, usually have a combined underground drainage installation within the curtilege of the property. These older installations may also have had an additional trapped access fitment which was prone to blockage: this is known as an interceptor trap. It was intended to prevent gases and rats from the sewer entering the house. They were largely unnecessary as the waste stack was usually connected to a trapped gully and the soil and vent stack had trapped WCs.

Combined drainage – 1930s example

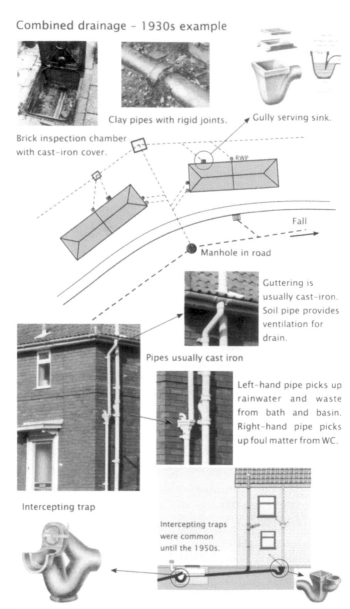

Clay pipes with rigid joints.

Gully serving sink.

Brick inspection chamber with cast-iron cover.

RWP

Fall

Manhole in road

Guttering is usually cast-iron. Soil pipe provides ventilation for drain.

Pipes usually cast iron

Left-hand pipe picks up rainwater and waste from bath and basin. Right-hand pipe picks up foul matter from WC.

Intercepting trap

Intercepting traps were common until the 1950s.

Separate drainage – modern example

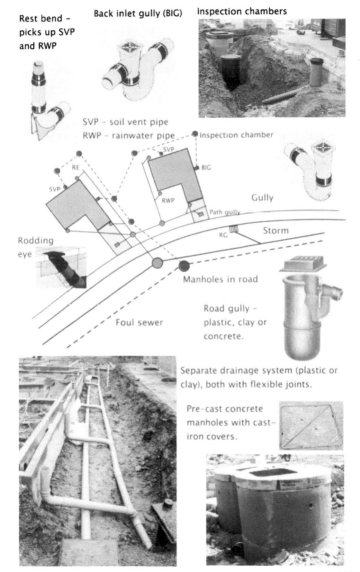

Rest bend – picks up SVP and RWP

Back inlet gully (BIG)

Inspection chambers

SVP – soil vent pipe
RWP – rainwater pipe

Inspection chamber

RE

SVP

SVP

BIG

RWP

Path gully

Gully

RG

Storm

Rodding eye

Manholes in road

Foul sewer

Road gully – plastic, clay or concrete.

Separate drainage system (plastic or clay), both with flexible joints.

Pre-cast concrete manholes with cast-iron covers.

Below-ground drainage generally has a long, relatively trouble-free life. It is, therefore, a little understood, and frequently ignored, aspect of building construction. Underground drainage systems should be laid to suitable gradients and in straight lines. The access arrangements are of particular importance in order to remove blockages. These not only stop the use of the drain itself, and create back-up, but also can cause more serious damage, leading to drain collapse or undermining of soils.

Access fittings range from the simple rodding eye (see previous diagram) to a manhole that enables access at drain level. Generally there should be access fittings on an underground drainage system at:
- any change of direction
- any change in gradient
- any junction of drain branches
- at specified distances required by the Building Regulations (to allow for rodding, jetting etc).

Materials

Above ground: Nowadays, the most common material is uPVC. This has the advantage of being relatively cost effective, corrosion resistant, light in weight and relatively simple to joint. However, it becomes brittle with exposure to ultraviolet radiation, it has limited strength against impact/mechanical damage and can sag if not adequately supported. It also moves significantly in relation to temperature. Cast-iron was the most popular material in the past and this has the advantage of good mechanical resistance, good longevity (if appropriately maintained) but it is heavy, expensive and requires regular maintenance (ie painting).

Under ground: uPVC has also become a popular choice for underground drainage systems. This material provides a degree of flexibility (useful if ground movement is expected), is relatively durable and requires fewer joints than the other materials. It can become brittle if exposed to UV radiation and is susceptible to crushing.

Salt-glazed and vitrified clay pipes have been used for many years and are still a popular choice. They have high strength and good resistance to corrosion. Nowadays they are fitted with flexible joints to allow for minor ground movement. They are heavier than plastic pipes but have less onerous bedding requirements.

Both materials require differing types of bedding and surround materials according to circumstances and will require different covering (ie trench back-fill) arrangements.

Above–ground drainage problems

Typical problems of two-pipe systems include:
- Blockage of hopper heads.
- Blockage of interceptors.
- Blockage of ventilation on the soil and vent pipe caused by birds nests/leaves. A wire cage should be fitted.
- Poor siting/detailing of top of soil and vent stack can lead to gases entering the house.
- Lack of trapped gully at the underground connection for the waste stack.
- Syphonage caused by poorly detailed waste branches or poorly formed traps.

Such systems are most likely to be cast-iron. Defects include:
- Rust and/or holes, particularly at the rear of the pipe – the part that is difficult to repaint.
- Joints in such pipes are formed with hemp/cement or sometimes lead. These can fail over time and allow leakage.
- Failure of, or insufficient fixings for, this heavy material.

Single–stack systems

Typical problems include:
- Failure to adhere to Building Regulations regarding minimum depth of trap, and length, diameter and gradient of waste branches (this can lead to siphonage).
- Location of WC branch too close to waste branch from bath, wash hand basin etc.
- Blockage of ventilation to the soil and vent pipe caused by birds' nests and leaves etc.

Gutters and downpipes

Typical problems include:
- Poorly fixed gutters (ie insufficient or missing gutter clips) or cast-iron gutters ineffectively fixed to fascia board.
- Undersized gutters.
- Inappropriately fixed gutters – too close or too low to capture rainwater discharging from the roof.
- Leaking joints on gutters: joints between sections should, ideally, allow for thermal movement. Cast-iron joints tend to be screwed together with a waterproofing compound between sections that can harden over time. In uPVC, push-fit joints with flexible gaskets can degrade over time. uPVC can also become brittle on exposure to ultra violet radiation. Plastics have a high rate of thermal movement – to avoid buckling the gutter sections should have a few millimetres of free play in them where they join.
- Clearance of gutters: debris from roof covering (particularly in concrete tile roofs), moss, guano and vegetation can quickly build up, encouraging vegetation growth (buddleia being a particular culprit). Clearance for inspection and maintenance are a sensible annual task.
- Cast-iron gutters require regular repainting with a bitumous paint internally to avoid long-term corrosion. Cast-iron also requires external decoration.
- Splits and damage will occur in downpipes if there are blockages and water freezes in the downpipe.
- Inadequate fixing of downpipes, leaking joints, missing brackets.

Below–ground drainage

Most old installations will not be completely watertight – so testing is not really worthwhile unless specific problems are suspected. The most common problem is drain blockage – usually caused by inappropriate materials being flushed down the toilet. Blockages can also be caused by subsidence, tree roots and traffic damage.

Other, more general, problems include:
- Lack of access fittings at changes in direction, gradient and meeting of branches makes the drainage system difficult to rod or jet, or even inspect.
- Rusty inspection chamber covers that cannot be lifted.
- Rigid clay pipe joints often cause cracking, resulting in leaks or pieces of fractured drain falling into the pipe.

Gutters and downpipes

Clockwise from top left. (1) This gutter has been fixed too low. When it rains, water shoots over the top of the gutter. (2) Splashing from the downpipe can ice in freezing weather, making the footpath treacherous. (3) In heavy rain the end of this gutter overflows because of the amount of water running down the valley. (4) This gutter leaks at the joints.

This cast-iron downpipe (bottom left) has been leaking for many years. Leaks can occur through failure of the joints or because of rust – usually due to lack of painting at the back of the pipe.

Gutters should be cleaned out regularly. Once vegetation takes a hold, there is a risk of overflowing, damp penetration and even rot in fascias and rafters.

Missing section of downpipe.

- Blockage of trapped gullies (leaves are often the main culprit).
- Lack of trapped gullies at the foot of waste stacks (two-pipe system).
- Tight bend at the foot of soil and vent stack can lead to blockage and back siphonage.
- Layouts are sometimes overly complex or have:
  - o branch drains running contra flow to main drain
  - o drains located too close or under trees
  - o drains laid under driveways/roads without adequate cover/bedding arrangements
  - o drains passing through substructure of house without provision for movement.

Drainage systems are best inspected while they are in operation: for example:

- Gutters and downpipes should be inspected when it is raining.
- When checking drainage above ground, fill the sink and bath etc. Look for leaks when plugs are pulled out and listen for gurgling sounds that might suggest siphonage. You can often smell 'broken' traps.
- When inspecting drainage below ground lift an inspection cover and ask someone to flush a WC – see how well and how quickly the water flows through the system.

CHECKLIST

Areas to inspect	Comments/problems
Gutters	Check that:

Gutters   Check that:
- There are no leaks or drips (particularly at joints).
- They are effectively sited so that rainwater is caught by the gutter and does not spill down the façade of the building.
- They are of adequate size and do not easily overflow.
- They are securely fixed with sufficient brackets and/or onto the fascia.
- They are not sagging through insufficient support over their length.
- The decorative state of the internal and external surfaces (of particular importance in cast-iron).

Areas to inspect	Comments/problems
	• There is no build up of debris and vegetation.
Rainwater downpipes	Check that: • They are securely fixed to the structure (externally) or that the boxing-in provides sufficient access (internally). • The joints are not leaking. • They are not blocked. • The decorative finishes are in good condition, particularly the rear of the pipe facing the wall (cast-iron). • If they discharge to a gully there is no risk of excess splashing.
Above-ground drainage	Soil and vent pipes, check: • They are effectively secured to the structure. • They extend beyond the eaves level (higher if there is a roof light). If they terminate within the structure of the building they terminate in a one-way, air admittance valve. • There is nothing blocking the ventilation, ie the SVP is open to the atmosphere. • There is wire cage protection to the top of the pipe. • There are no leaks. • The branch waste for the WC is at least 200mm above any other branch waste. • The decorative condition (cast-iron) particularly the rear of the pipe facing the wall. Hopper heads, check: • For secure fixing to structure. • Wire cover to avoid leaves and other external debris entering the hopper head (these are quite rare). • Decorative condition (if cast-iron). Waste stacks and branch waste pipes, check for: • Leaks, particularly at joints. • Sufficient support brackets and a lack of sagging, which can disrupt efficient flow in the branch.

Areas to inspect	Comments/problems
	• Appropriate diameter of pipe, gradient and length of branch for appliance (in accordance with Building Regulations). • Decorative condition (if cast-iron). Under sinks etc (where possible) check for: • Presence of suitable trap. • Lack of leaks, particularly at joints.
Below-ground drainage	Checking anything underground is difficult. It might be worth locating and lifting all the inspection chamber covers to check the inside of the chambers. It might also be worth observing the effectiveness of the drainage run as a WC is flushed and taps are run. It might also be worth trying to ascertain how, and where, the drains run. In other words, is the system combined or separate, does the system run out into the pavement or road, are drains shared with neighbours etc.

ROT AND INSECT ATTACK

TIMBER VULNERABILITIES

In modern construction, timber, or increasingly its derivatives, form the roofs, floors and (sometimes) the wall structures of many houses. Timber remains the most popular material for windows, doors, stairs and many other more decorative elements within houses. Yet timber has some significant vulnerabilities that can lead to very serious and expensive defects: fungal infestations and insect attack can both destroy timber. Controlling the level of moisture is a prerequisite to breaking the life cycle of the fungi and insects that feed on timber.

Until the 17th century, the majority of timber used in houses was native hardwood, such as oak. Since the 19th century, imported softwoods have been more common. Softwoods are generally less durable and softer than hardwoods (although there are some significant exceptions), however the distinction between hard and softwoods is botanical. Softwoods are from trees that are evergreen and thus do not shed their leaves during the winter period. Hardwoods are derived from deciduous trees that shed their leaves during a winter dormant period.

Whatever their botanical classification, trees are effectively factories for the production of structural and consumable sugars. The process of photosynthesis converts water, drawn up by the roots of tree, carbon dioxide, absorbed from the air, into complex structural carbohydrates – sugars. The energy for this process is provided by sunlight. Each year's growth consists of a layer added to the outside of the stem (trunk and all the branches) just underneath the bark. The first few layers of more recent growth rings are known as the sapwood. This is where sugars and water are distributed around the tree. As the tree ages and new annual growth rings are added, the inner part of the trunk dies, stops transporting sugars and moisture, darkens and becomes more dense. This is known as the heartwood and, as the tree ages, becomes an increasing proportion of the trunk of the tree. In terms of durability, the sapwood, being less dense and containing less sugars, has very little durability. The inherent durability of the timber is retained in the heartwood.

Over 90% of softwoods are imported because the UK's climate is too temperate to produce dense and durable softwood timber. Since the end of the First World War, and particularly since 1945, worldwide demand for timber has grown and production methods have responded by the creation of plantations (as opposed to wild forests), changing to fast-growing tree species, and by significantly reducing the growth period of the trees. Thus, timber in a Victorian house may well have been from ancient forests and grown for several centuries: the proportion, therefore, of durable heartwood to non-durable sapwood was high. In a modern house, softwood may only be 30–40 years old. This means that softwood used in buildings today has a far higher proportion of sapwood with significantly less natural durability than timber used pre-1945. Consequently, it is much more vulnerable to fungal and insect attack.

Fungal attack

Fungi have no chlorophyll and are unable to produce food through photosynthesis. They feed by attacking the sugar in timber and converting it into a soluble food. The most significant effect of a fungal attack is loss of strength. Infested timber will crack, soften, lose weight and change colour. Fungal attack is also known to 'soften-up' timber, making insect attack far more likely.

There are four key environmental conditions enabling fungal infestation:

- Moisture is the most critical factor: without moisture fungi cannot grow or reproduce. At a moisture content of less than 20% or so, fungal infestation is unlikely to occur.
- Food source. Sapwood is far more susceptible to fungal infestation; heartwood is not immune.
- Oxygen is required by fungi.
- Extremes of temperature can stop or retard the life cycle of fungi.

If there is a major change in environmental conditions the life cycle can be broken or, more dangerously, can make the fungus dormant.

Two types of fungal attack can occur, which are commonly known as wet rot and dry rot. While this definition is technically incorrect, it is a very important distinction as the nature, the effect and, more importantly, the impact of the effect of these two types is very different. The scale of damage from a dry rot outbreak is the greater threat.

Dry rot is known as such because of its effect on the timber – it dries and cracks into 'cuboid' shapes (shown below), becomes crumbly and loses strength. The conditions for propagation of spores are:

- still air of high humidity
- moisture content of timber of 20% or more
- temperatures of 5–40°C.

Once germination of the fungal spores occurs, the hyphae (white to grey strands) and mycelium (a cotton-wool like mass with pale yellow and purple patches) can spread rapidly. As they spread, they attack and break down suitably damp timber. Strands within the mycelium can transport both food and moisture to the active growth area.

A dry rot outbreak is particularly serious as it can travel over inert surfaces, such as masonry and plasterwork, in the search for food sources and suitable conditions to continue its life cycle. The most common location for dry rot attacks is poorly ventilated situations where timber is, and remains, damp for periods of time, ie unventilated cellars, ground-floor and sub-floor voids.

Wet rot

This requires much more moisture than dry rot and commonly attacks saturated timbers (areas liable to frequent flooding or in areas with concealed plumbing leaks). Some forms of wet rot can be mistaken for dry rot as the distinctive cuboidal cracking results in decay of a similar nature. However, unlike dry rot, infestations of wet rot attacks tend to be isolated. Other forms of wet rot affect timber so it becomes lighter in colour and weight and develops a fibrous lint-like consistency with cracking along, rather than across, the grain. Typically such an attack will be observed in poor-quality softwood windows where the decorative finish has not been maintained.

Insect attack

For some insects, specifically a number of members of the beetle family such as the Common Furniture Beetle, timber is their primary food source and means of shelter.

The life cycle of wood-boring beetles is as follows:

- **Eggs** are laid by the mature female in cracks and other fissures in the timber. In the right circumstances (timber with a high moisture content is ideal), these will hatch into larvae.
- The **larvae stage** is the most destructive to timber as the larvae burrow into the timber in search of food and shelter. There is no sign damage done as this takes place within the timber itself. With the Common Furniture Beetle, this stage can last for two to four years.
- **Pupa and adult stage.** If the environmental conditions support the larvae it will eventually cease to feed, metamorphose into a winged insect, bore its way back out of the timber, mate and die.

The flight holes, which are the means of identification of an 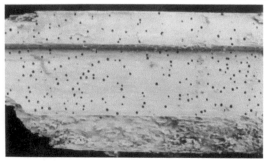 infestation, are created as the adult emerges. Attacks can be very localised, in response to food quality, humidity and temperature. The better the conditions, the quicker the life cycle of the beetles is. The Common Furniture Beetle prefers sapwood but will attack heartwood softened-up by fungal attack. The adult beetle is 3–5mm long and it emerges from timber between May and August. Infestations are likely in any locations where timber can become more moist, eg roof spaces and timbers under ground-floor voids. The image above shows a heavy beetle infestation.

Approaches to treatment

With both fungal and insect attack, the key strategy must be prevention not cure. Much modern softwood is routinely pre-treated with chemicals that poison the timber and thus render the timber useless as a food source. Such treatments, prior to installation, are far more effective than post-construction, *in situ* treatment.

In the past, a number of chemicals were used that appeared to be as, if not more, harmful to the human occupants of the treated building, than to the infestations themselves. Some approaches were not only dangerous to operatives and occupants, but also they have been found to be ineffective. Approaches in the recent past not only included precautionary chemical treatments, they also included precautionary removal of unaffected timbers. As modern replacement timber is likely to have significantly less natural durability than earlier timber, such approaches are increasingly seen as counter-productive. There is an emerging consensus favouring biological rather than biochemical approaches to both fungal and insect attack. Understanding the conditions that enable a successful life cycle for fungi and wood-boring insects, certainty about the form of infestation, undertaking basic maintenance focused on ensuring appropriate and sustained moisture content and vigilance of potential infestation locations, are key factors in preventing fungal and insect attack.

CHECKLIST

This checklist differs from other sections because the correct identification of insect and fungal infestations is critical and requires expert, and correct, diagnosis: misdiagnosis, particularly of dry rot, can be disastrous. This checklist identifies key areas in houses that are most susceptible and broadly describes what evidence to look for. If such evidence is found specialist advice is essential.

- **Understand which areas of the house are most susceptible.**
  This will include areas that:
  1. Have timber built into them: mostly these areas are obvious (eg roofs, floors, wall panelling etc). Some are less obvious, for example old solid walls with timbers built into the brickwork or stonework.
  2. Are at risk from saturation or dampness.
  3. Have low ventilation. This will decrease the rate of evaporation and increase the moisture content of timbers.

Typically, the areas where both fungal and insect infestations are likely to appear are the relatively inaccessible damp, warm areas of a house. Most insect and fungal infestations will remain localised where conditions support their life cycle. However, dry rot has the ability to rapidly travel long distances in the search for suitable levels of moisture and food to sustain its life cycle.

- **Typical areas to investigate:**

  Suspended timber ground floors. Early (pre-1920) practice did not always include DPCs, had no concrete oversite separating moisture-laden ground from the timbers of the floor and had inadequate ventilation.

  Timber roof structures. Particularly those without effective and suitable ventilation are at risk when condensation or roof covering leaks occur. Valley and parapet gutters can suffer minor damage to the waterproofing membrane (most frequently lead sheet supported by timber boarding), which creates chronic leakage. Ineffective flashings, particularly the complex ones around chimneys can also lead to chronic leakage.

  Poorly maintained external timber joinery. The inferior nature of post-1945 joinery has led to frequent problems of wet rot. 1970s softwood windows are very vulnerbale.

  Areas of the building located adjacent to rainwater goods and soil stacks. Such pipes can leak, become blocked or can be damaged producing long-term but, perhaps, not obvious penetrating damp. These can be particularly dangerous to upper-floor structures in solid masonry (non-cavity) construction and can cause infestations affecting all the floors and related timber structures (such as timber stud partitions) within the house.

  Areas immediately behind WCs and areas adjacent to internal waste pipes. The connection between the soil pipe and the WC is, nowadays, formed with a flexible rubber joint. This can fail or be disrupted and cause leaks. Poorly maintained or damaged plumbing and waster waste pipes can also cause long-term leaks.

- **Signs of infestations – what to look for:**

  Insect infestation – look for flight holes. Identification of current infestation is vital. Look for holes with fresh frass or pellets. Note that new flight holes are likely to occur between spring and summer.

  Fungal infestations – look for disrupted surface finishes, for example cuboidal cracking in skirting is a common indicator. Sometimes spores, which look like a reddish dust, appearing regularly on surfaces are a sign of the proximity of the fruiting body. In some circumstances, parts of the fungus itself will be encountered. Most fungi also produce a characteristic smell, which, in an enclosed space, will be obvious.

The two photos below show evidence and consequences of dry rot.

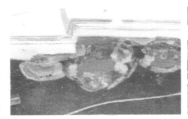 

INDEX

T - #0390 - 071024 - C294 - 178/108/14 - PB - 9780728204898 - Gloss Lamination